REQUEST FOR REPROGRAPHICS

DERBY College

Normally **2 working days** notice is required. Please ensure all sections of the form are fully completed as requests may be subject to delay whilst clarification is sought. Wherever possible, copy quantities should be kept to a minimum to prevent unnecessary waste of paper and technicians' time due to later amendments or over ordering.

Name: HONG XU	Site/Room No J.W.C	
Department : MATHS	Date: 7/9/0	Tel: 5703

No. of originals:	No of copies:	Latest date required:

A4 Paper

				A4 Card		**Size**	**A3**
White ☐	Deep Green ☐			White ☐			
Pale Yellow ☐	Deep Red ☐			Pale Yellow ☐			
Pale Blue ☐	Deep Yellow ☐			Pale Blue ☐			
Pale Cream ☐	Lime Green ☐			Pale Pink ☐		**A4**	**A5**
Pale Salmon ☐	Lilac ☐			Pale Green ☐			
Pale Pink ☐	Grey ☐			Bright Blue ☐			
Pale Gold ☐	Pale Green ☐			Deep Green ☐			
Bright Blue ☐				Deep Yellow ☐			
				Deep Red ☐			

SPECIAL INSTRUCTIONS

Back to back ☐
Stapled ☐
Collated ☐
Spiral Bound ☐

Name:	To be collected ☐
Site: JWC Room: F26	Postroom Block (If PCA)	To be returned ☑

AQA Mathematics for GCSE

Exclusively endorsed and approved by AQA

Homework Book

Series Editor
Paul Metcalf

Series Advisor
David Hodgson

Lead Author
Margaret Thornton

June Haighton
Anne Haworth
Janice Johns
Steven Lomax
Andrew Manning
Kathryn Scott
Chris Sherrington
Mark Willis

FOUNDATION
Linear 1

Nelson Thornes
a Wolters Kluwer business

Published in 2006 by:
Nelson Thornes Ltd
Delta Place
27 Bath Road
CHELTENHAM
GL53 7TH
United Kingdom

06 07 08 09 10 / 10 9 8 7 6 5 4 3 2 1

A catalogue record for this book is available from the British Library.

ISBN 0 7487 9777 7

Cover photograph: Salmon by Kyle Krause/Index Stock/OSF/Photolibrary
Page make-up by MCS Publishing Services Ltd, Salisbury, Wiltshire

Printed and bound in Spain by GraphyCems

Acknowledgements

The authors and publishers wish to thank the following for their contribution:
David Bowles for providing the Assess questions
David Hodgson for reviewing draft manuscripts

Thank you to the following schools:
Little Heath School, Reading
The Kingswinford School, Dudley
Thorne Grammar School, Doncaster

The publishers thank the following for permission to reproduce copyright material:

Vitruvian Man – Corel 481 (NT): p. 97

The publishers have made every effort to contact copyright holders but apologise if any have been overlooked.

Contents

Introduction

This book contains homework that allows you to practise what you have just learned. Each chapter is divided into sections that correspond to the numbered Learn topics for the matching chapter in the Students' Book.

 Means that these questions should be attempted with a calculator.

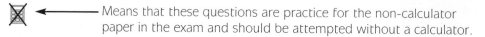 Means that these questions are practice for the non-calculator paper in the exam and should be attempted without a calculator.

1 ◄——— Underlined questions are harder questions.

Coursework

This section explains the coursework mark scheme and features three Handling data mini-coursework tasks.

1 Statistical measures

Homework 1

1 Find the mean and range of each set of numbers.

   a 2 5 7 11 15

   b 4 7 9 15 17 5 5 12 10 7

2 Find the mode and median of each set of numbers.

   a 4 5 7 7 7 8 9 10 11 12 12 12 12 13 13

   b 3 5 1 6 5 2 8 4 1 5 5 2 6 3

3 For this set of numbers

   11 8 5 7 3 2 7 7 3 4

   find: **a** the mean   **b** the mode   **c** the median.

4 A chart in the Health Centre shows that the mean weight of a 6-month old baby is 7.5 kg. There are six babies at the Health Centre. Their weights are:

   Katie 7 kg   Sarah 8 kg   Chris 6.5 kg   Richard 7.5 kg

   Natalie 6 kg   Sam 8.5 kg

   a What is the range of the babies' weights?

   b What is the median weight?

   c How many babies are over 7.5 kg?

   d What is the mean weight of the six babies?

5 Emily is training for the British Orienteering Championships. Her coach says that her training should average not less than 30 minutes running a day.
So from Monday to Friday she timed how long she ran.

   Here are the results: 20, 32, 28, 25, 45 minutes.

   a What is the range of her times?

   b Calculate her mean training time.

   c Did she reach her coach's target?

6 Nick is keeping a record of his golf scores. In three rounds he scored 85, 85 and 82.

On his next round he scored 100.

a Calculate his mean score for the four rounds.

b To enter the club championship he needs a mean score of 89 or less. Does he qualify?

7 The mean of three numbers is 21. Two numbers are smaller than the mean and one is bigger. Write down three possible numbers.

8 Grace is checking her module tests. She scored 65 in her first module, 75 in her second module and 85 in her third module.

She is about to take her last module. She needs to get an average of 75 or more to get a grade A.

Work out the least score she needs to get on her last module.

9 In a diving competition, Tom scores a mean mark of 5.3
Seven of his eight marks were   4.9,   5.3,   5.5,   5.6,   5.8,   4.8,   4.9
What was his eighth mark?

Homework 2

1 Write down the time you spend on each of these activities each week.

a sleeping   **b** homework   **c** watching television

Calculate the mean number of hours for each and write your answers to the nearest hour.

2 In a survey on the number of people in a household the following information was collected from 50 houses.

Number of people in a household	Number of households
1	9
2	19
3	9
4	8
5	4
6	1
Total	50

a Find the mean, median, mode and range of household sizes.

b Which average is the best one to use to represent the data? Explain your answer.

3 The graph shows the frequency of goals scored by a football team during a season.

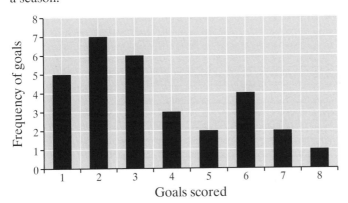

a How many matches were played during the season?

b What is the modal score?

c What is the median number of goals scored?

d What is the mean score?

e What is the range of the number of goals scored?

4 Miss Saunders divided her class of 40 students into two groups, A and B. She gave each group the same test, which was marked out of 10. The results are shown in the table below.

Number of marks	0	1	2	3	4	5	6	7	8	9	10
Group A frequency	3	0	0	5	1	6	1	2	0	0	0
Group B frequency	0	0	0	3	3	3	6	4	1	1	1

Which group did better?
Compare the two groups by finding the range and mean scores.

5 This bar chart shows the number of letters in words in a crossword puzzle.

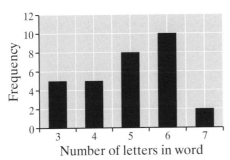

Find the:

a median   b mode   c mean   d range of the number of letters.

Which average best represents the distribution?
Give a reason for your answer.

Homework 3

1   This table gives the number of years' service by 50 teachers at the Clare School.

Number of years' service	Number of teachers
0–4	11
5–9	15
10–14	4
15–19	10
20–24	6
25–30	4

a Find the modal class.

b Calculate an estimate of the mean.

2   In a science lesson 30 runner bean plants were measured. Here are the results correct to the nearest centimetre.

6.2	5.4	8.9	12.1	6.5	9.3	7.2	12.7	10.2	5.4
7.7	9.5	11.1	8.6	7.0	13.5	12.7	5.6	15.4	12.3
13.4	9.5	6.7	8.6	9.1	11.5	14.2	13.5	8.8	9.7

The teacher suggested that the data was put into groups.

Length in centimetres	Tally	Total
5 but less than 7		
7 but less than 9		
9 but less than 11		
11 but less than 13		
13 but less than 15		
15 but less than 17		

a Copy and complete the table.

b Use the information to work out an estimate of the mean height of the plants.

c Calculate the mean from the original data.

d Why is your answer to part b only an estimate of the mean?

3   The heights of 50 students at Eastham School were measured. The results were put into a table.

Height (h cm)	Frequency
$149.5 \leqslant h < 154.5$	4
$154.5 \leqslant h < 159.5$	21
$159.5 \leqslant h < 164.5$	18
$164.5 \leqslant h < 169.5$	7

Estimate the mean value of the distribution.

2 Angles

1 How many degrees are there in

   a two full turns   **b** one sixth of a turn   **c** one eighth of a turn?

2 What fraction of a turn is

   a 90°   **b** 30°   **c** 210° ?

3 Anna faces south and makes a $\frac{1}{4}$ turn clockwise. Which way is she facing now?

4 What angle does the hour hand of a clock move through between 5 o'clock and 7 o'clock?

5 What is the angle between the hour hand and the minute hand at 5 o'clock?

6 What is the angle between the hour hand and the minute hand at half past three?

Homework 2

Copy this table. Put a tick in the correct column for each angle.

Angle	Acute	Obtuse	Reflex
102°			
85°			
203°			
194°			
315°			
28°			
177°			
99°			

Homework 3

1   Calculate the size of each of the marked angles.

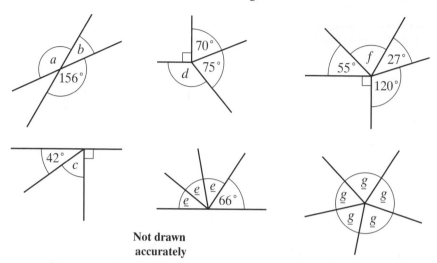

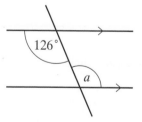

**Not drawn
accurately**

Homework 4

In questions **1** to **6**, work out the size of each of the marked angles.
Write down a reason for each step of your working.

1

3

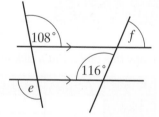

2

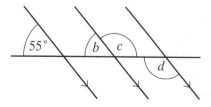

4

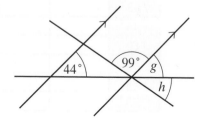

**Not drawn
accurately**

5
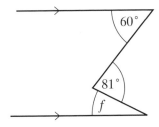

6
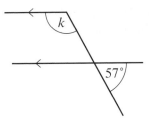

Explain why the answers given in questions **7** and **8** are wrong.

7
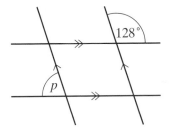

Gary says that $p = 72°$

Not drawn accurately

8
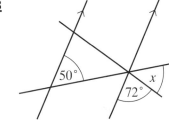

Parveen says that $x = 48°$

Homework 5

1 For each diagram, write down the three-figure bearing of P from Q.

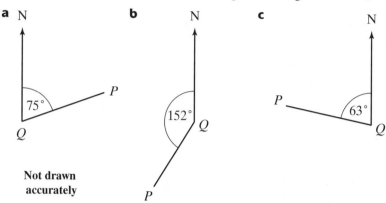

a N

75°

Q

P

**Not drawn
accurately**

b   N

152°

Q

P

c   N

P

63°

Q

2 For each of the diagrams above, write down the bearing of Q from P.

3 Draw a sketch to show each of these:

   a the bearing of G from F is 094°

   b the bearing of L from K is 265°

   c the bearing of V from U is 310°.

3 Integers

1   Put these numbers in order, smallest first.

   a $+3, -2, +4, -6, -5, +7, -1, 0, +5$

   b $-3, +2, +5, -7, -1, +7, +3, -6, -8, -2$

   c $+5, -9, -2, +6, -5, +84, -87, +62$

2   In Leeds the temperature was $-4°C$. In Southampton, on the same night, the temperature was $5°C$. Which city was warmer?

3   Choose the correct sign, $<$, $>$ or $=$, to go between these numbers.

   a $+8$ $+5$

   b -3 $+2$

   c 2 -2

   d -4 -5

4   The water level in a reservoir is described as $+2$ if it is 2 metres above normal and -2 if it is 2 metres below normal.
   What do the following mean?

   a -5 b $+1$ c 0

5   Natalie chooses a number and then describes it to her friends.

   'It is an integer. It is between -3 and -6. It is less than -4.'

   What is the number?

6   Antony chooses a number and then describes it to his friends.

   'It is an integer. It is between -4 and $+2$. It is between -10 and -2.'

   What is the number?

7   Make up five number descriptions like Natalie's and Antony's.
   Make sure that negative numbers are included.

Homework 2 ⊠

1   Find the missing numbers in each of the following.

   a $-3 + \ldots = -2$ **c** $3 - \ldots = +8$ **e** $8 - \ldots = -3$

   b $3 + \ldots = -2$ **d** $5 + \ldots = 2$ **f** $-4 - \ldots = +2$

2   The answer is -3. Write down five addition sums with this answer.

3   The answer is -4. Write down five subtraction sums with this answer.

4   **a** The temperature is 4°C. It rises by 5°C. What is the new temperature?

   b The temperature is 3°C. It drops by 4°C. What is the new temperature?

   c The temperature is -2°C. It rises by 6°C. What is the new temperature?

   d The temperature is -6°C. It drops by 3°C. What is the new temperature?

5   Copy and complete this addition table.

   Some answers have been done for you.

					B			
	$A + B$	-3	-1	-5	$+2$	$+5$	$+1$
	-4	-7					
	-2				0		
A	$+2$						
	$+6$						
	-3			-8			
	$+1$						

6   Copy and complete this subtraction table.

   Some answers have been done for you.

					B			
	$A - B$	-5	-1	$+2$	-3	$+6$	$+7$
	$+3$	$+8$					
	-1				$+2$		
A	$+1$						
	$+5$						
	-2				-4		
	-3						

Homework 3

1 Find the missing number in each of the following.

   a $+4 \times \ldots = +16$ **c** $-4 \times \ldots = -8$ **e** $-8 \div \ldots = -4$

   b $+2 \times \ldots = -10$ **d** $-2 \times \ldots = +4$ **f** $\ldots \div -4 = 8$

2 Copy and complete this multiplication table.

   Some answers have been done for you.

B

$A \times B$	-5	-1	$+2$	-3	$+6$	$+7$
+3	-15					
−1				$+3$		
+1						
+5						
−2			-4			
−3						-21

A

3 The answer is -16. Write down four multiplication sums with this answer.

4 The answer is -2. Write down four division sums with this answer.

5 Work out:

   a $\dfrac{-2 \times +6}{-3}$ **b** $\dfrac{-4 \times -9}{2}$ **c** $\dfrac{8 \times -6}{-4}$

6 Find two numbers whose:

   a sum is -7 and product is 10

   b sum is 5 and product is -6

   c sum is -8 and product is 12

   d sum is -2 and product is -15.

Homework 4

1 Write all the factors of these numbers:

   a 16   **b** 24   **c** 60   **d** 100

2 Write the first five multiples of these numbers:

   a 4   **b** 10   **c** 14

3 Find the least common multiple (LCM) of each set of numbers.

   a 2 and 7   **c** 4 and 5   **e** 4, 6 and 9

   b 5 and 3   **d** 8 and 5

4 Find the highest common factor (HCF) of each set of numbers.

   a 8 and 12   **c** 12 and 18   **e** 48 and 72

   b 8 and 18   **d** 24 and 30

5 Look at these numbers:

   2, 3, 4, 5, 6, 7, 8, 9, 10, 11, 12, 13, 14, 15

   a Which of them are factors of 12?

   b Which of them are multiples of 6?

   c Which of them are factors of 15?

   d Which of them are multiples of 4?

6 **Get Real!**

   Two buses stop at the same bus stop.
   On one route the bus stops every 25 minutes.
   On the other route the bus stops every 40 minutes.
   If both buses leave the stop at 9 a.m., at what time should the buses next
   reach the stop together?

7 One light flashes every 8 seconds.
   A second light flashes every 10 seconds and a third every 15 seconds.
   If the three lights start flashing at the same time, after how long will they
   flash together again?

Homework 5

1 Write the following numbers as products of prime factors.

   a 52   **b** 60   **c** 24   **d** 64   **e** 80

2 Express each of these numbers as a product of its prime factors.
Write your answers using index notation.

   a 70   **b** 96   **c** 100   **d** 150   **e** 256

3 Write 72 as a product of prime factors.

4 p and q are prime numbers. Find the values of p and q when $p^3 \times q = 24$.

5 What numbers are these?

   a It is a prime number. It is a factor of 21. It is not a factor of 12.

   b It is a prime number. It is a factor of 132. It is bigger than 5.

Homework 6

Apart from question 6, this is a non-calculator exercise.

1 Find the reciprocal of each of these numbers.

   a 2   **b** 9   **c** 6   **d** 15

2 Find the reciprocal of each of these numbers.

   a $\frac{1}{3}$ **b** $\frac{1}{5}$ **c** $\frac{1}{11}$ **d** $\frac{2}{3}$

3 What is the value of $1 \div \frac{1}{4}$?

4 Find the reciprocal of each of these numbers.

   a 0.2   **b** 0.75   **c** 0.6   **d** $0.\dot{6}$

5 Find the reciprocal of each of these numbers.

   a $3\frac{1}{5}$ **b** $6\frac{7}{8}$ **c** $2\frac{2}{5}$ **d** $4\frac{9}{10}$

6 Find the reciprocals of the numbers from 30 to 40. Write them correct to 4 decimal places if they are not exact decimals. Which of the numbers have reciprocals that are:

a exact decimals

b decimals with one recurring figure

c decimals with two recurring figures

d decimals with three recurring figures?

4 Rounding

1 Round these numbers to the nearest whole number.

   a 12.6   **b** 3.6   **c** 103.9   **d** 99.6   **e** 0.8   **f** 0.4

2 Round these numbers to

   a the nearest 10

   b the nearest 100.

   i 236   **ii** 881   **iii** 568   **iv** 2045   **v** 24.8

3 Round the numbers in question **2** to the nearest 5.

4 Round these numbers to the nearest thousand.

   a 24 556   **b** 3406   **c** 578 195

   d three hundred and fifty-four thousand two hundred and fifty-six.

5 Explain why 96 756 does not become 97 when rounded to the nearest thousand.

6 What is the smallest number that, when rounded to the nearest whole number, becomes 10? What is the largest?

7 Write down four different three-digit numbers that become

   a 270 when rounded to the nearest 10

   b 270 when rounded to the nearest 5.

   c Which of your numbers could be the answers to both parts **a** and **b**?

8 Draw a picture to show the possible lengths that a pencil could have for its length to be 8 cm correct to the nearest centimetre.

9 Jan's height is 146 cm. What is this to the nearest metre? Nearest 10 cm? Nearest 5 cm?

10 Round these amounts of money to

    a the nearest pound

    b the nearest 50 pence.

    i £26.47   **iii** £405.89   <u>**v**</u> £0.85

    ii £3.63   <u>**iv**</u> 45 pence

<u>**11**</u> An amount of money is £63.50 correct to the nearest 50 pence.

    What is the largest amount it could be? What is the smallest?

12 A school has 1250 students, correct to the nearest 10 students. What are the smallest and the largest possible numbers of students in the school?

13 The attendance at a football match was 25 689. Write a sentence giving the approximate number of people at the match as if you were writing an article about the match for the local paper.

Homework 2

1 Round these numbers to one significant figure.

    a 326   **b** 589   **c** 3245

    Round these numbers to two significant figures.

    <u>**d**</u> 9999   <u>**e**</u> 9099   **f** 9950

2 Round these numbers to one significant figure.

    a 4.826   **c** 0.04826   **d** 0.004826

    b 0.4826

    Round these numbers to two significant figures.

    <u>**e**</u> 0.0004826   **f** 0.00004826

<u>**3**</u> Round the numbers in question **2** to two decimal places.

4 Get Real!

    Round these amounts of money to 2 decimal places
(that is, to the nearest penny).

    a £34.657   <u>**d**</u> £28.999

    b £33.490   <u>**e**</u> £0.3769

    c £435.672

5 Round these measurements to 1 decimal place
(that is, to the nearest millimetre).

   a 18.67 cm   **c** 68.23 cm   <u>**e**</u> 0.4545 cm

   b 8.38 cm   <u>**d**</u> 0.678 cm

<u>**6**</u> Round these masses to 3 decimal places
(that is, to the nearest gram).

   a 1.7683 kg   **c** 8.9247 kg   <u>**e**</u> 0.00035679 kg

   b 48.2467 kg   <u>**d**</u> 0.052905 kg

Homework 3

Do not use your calculator for estimating. You may, though, want to check your answers with the help of your calculator.

1 For each question, decide which estimate is best.

		Estimate A	Estimate B	Estimate C
a	3.67×7.4	2.8	2.1	28
b	1.8×21.4	4	40	400
c	$12.34 \div 4.1$	3	0.3	48
d	51.5×9.8	5	50	500
e	$6.1 \div 2.1$	0.3	3	30
f	44.1×1.8	80	44	440
g	$6.8 \div 1.1$	7	70	0.7
h	$2.8 \times 4.2 \div 3.8$	3	2	1
i	$31.9 \div 6.2 \times 3.2$	2	150	15

2 For each pair of calculations, estimate the answers to help you decide which you think will have the larger answer.

   a 4.2×2.8 or 4.1×2.75 **c** $13.41 - 4.28$ or $27.5 \div 2.14$

   b $33.3 \div 6.2$ or 2.31×2.78 **d** 59.4×31.6 or 95.3×9.8

3 Estimate the answers to these calculations by rounding to one significant figure.

a $8.6 + 4.9$

e 0.2×6

i $\dfrac{89.8 \times 5.9}{0.32}$

m $21.3(7.56 - 3.89)$

b $7.9 \div 2.1$

f $\dfrac{52.1 \times 4.7}{4.8}$

j $21.3(7.56 + 3.89)$

c $59.1 + 40.8$

g $\dfrac{28.5 + 53}{64.1 - 53.7}$

k $\dfrac{38.2 \times 7.9}{0.48}$

d 9.1×8.8

h $\dfrac{3.6 \times 7.1}{0.11}$

l $\dfrac{89.8 + 5.9}{0.32}$

4 **a** Write down two consecutive whole numbers, one that is smaller than the square root of 40 and one that is larger.

   b Write down two consecutive whole numbers, one that is smaller than the square root of 120 and one that is larger.

5 Estimate $6.32 + 4.26$ by rounding to the nearest whole number.

   Explain why the answer is an underestimate of the exact answer.

6 A group of 29 people weigh a total of 2772 kg.

   Estimate their mean weight.

   Is this a reasonable amount for the mean weight of a large number of people?

7 Estimate the total cost of 58 computers at £885 each.

8 Estimate the square roots of these numbers.

   a 15   **b** 30   **c** 60   **d** 5   **e** 2000

9 Estimate 4.13×8.22 by rounding to one significant figure.

   Explain why the answer is an underestimate of the exact answer.

10 Joe estimated the answer to $12.2 \div 8.2$ as 1.5 by rounding both numbers down. Does this mean that 1.5 is an underestimate of the correct answer?

11 Explain why $2.4 \div 0.6$ has the same answer as $24 \div 6$ and $240 \div 60$.

12 Which of these make a number larger and which of them make it smaller?

    a Dividing by 10

    b Multiplying by 10

    c Dividing by 0.1

    d Multiplying by 0.1

13 Estimate the answers to:

    a $41 \div 8.2$ **b** $41 \div 0.82$ **c** $41 \div 0.082$

14 **a** Find five numbers whose square roots are between 9 and 10.

    b Find two consecutive whole numbers to complete this statement:

      'The square root of 70 is between ... and ... '

15 **Get Real!**
Estimate the total cost of five CDs costing £6.98 each.

16 Estimate how many bottles of water costing 47 pence each you can buy for £8.

17 Estimate how much fencing is needed for a rectangular field measuring 48.5 m by 63 m.

18 A group of nine people win £4855 on the lottery.
Estimate how much each person will get when the money is shared out equally.

19 Matt drives at an average speed of 57.8 miles an hour for 2 hours 55 minutes.
Estimate how far he goes.

Homework 4

1 Each of these quantities is rounded to the nearest whole number of units. Write down the minimum and maximum possible size of each quantity.

    a 26 g   **c** 225 m   **e** 33 kg

    b 4 cm   **d** 13 litres   **f** £249

2   The upper bound of a length measured to 3 significant figures is 3.295 m.
If the actual length is x metres, copy and complete this statement:
... $\leqslant x <$ ...

3   A packet weighs 2 kg, correct to the nearest 100 g.
What is the maximum possible weight?

4   The amount of breakfast cereal in a packet should be 510 g, to the
nearest 10 g.
One packet weighs 504.6 g.
Tom says this is within the acceptable limits.
Explain why Tom is wrong.

5   The weight of a toffee is 5 g correct to the nearest half gram.
What is the minimum possible weight of one toffee?

Homework 1

1 Work out the angle marked by each letter and state what type of angle it is:

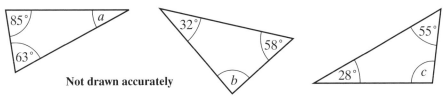

Not drawn accurately

2 Get Real!

The sketch shows a ladder against the side of a house.

Calculate the angle marked *x*.

Not drawn accurately

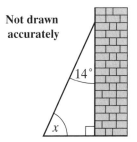

3 In triangle PQR, angle P is 90° and angle Q is 26°.
Sketch the triangle and calculate angle R.

4 Calculate the angles marked by letters.

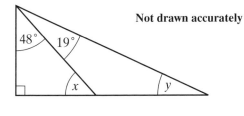

Not drawn accurately

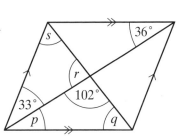

5 Triangle ABC has a right angle at A and angle B is 20° more than angle C.

Calculate angle B and angle C.

Homework 2

1 Work out the angles marked by letters.

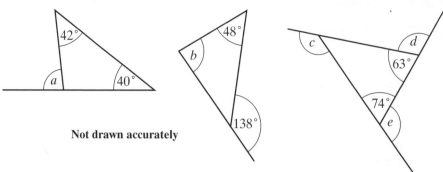

Not drawn accurately

2 Get Real!

The diagram shows the end view of a shelf.

Calculate the angle marked *x*.

Not drawn accurately

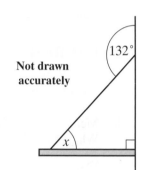

3 Calculate the angles marked by letters.

Not drawn accurately

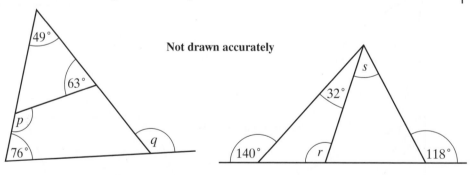

4 Get Real!

The diagram shows the positions of three buoys, A, B and C.

B is due east of A and C is due north of A.

The bearing of B from C is 131°.

a Calculate the angle marked *x*.

b What is the bearing of C from B?

Not drawn accurately

Homework 3

1 Work out the angles marked by letters.

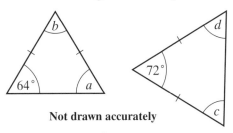

Not drawn accurately

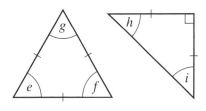

2 Get Real!

The diagram shows a cross-legged stool.

Calculate the angles marked by letters.

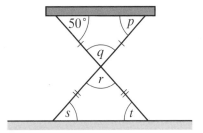

3 Measure the sides and angles of each triangle.
Which two triangles are congruent?

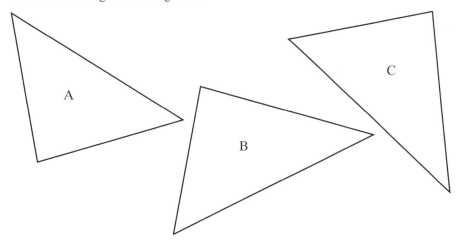

4 Calculate the angles marked by letters.

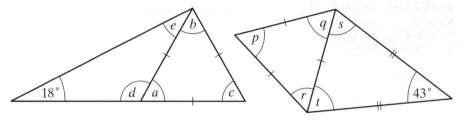

Not drawn accurately

5 One angle of an isosceles triangle is 35°.

Draw two possible triangles and give their other angles.

6 Use of symbols

1 Simplify:

   a $p + 2p + 3p$ **d** $p + 4p - 3p$ **g** $2ab + 3ab$

   b $t + t + 4t$ **e** $f + 6f - 10f$ **h** $5ad - 2da$

   c $d + 4d - 2d$ **f** $h + h - 5h + 2h$ **i** $p^2 + p^2 + p^2$

2 Simplify:

   a $x + 3x + 5 + x$ **d** $g + 2g + h$ **g** $7ab + 2b + 4a + 2ab$

   b $y + 2 + y + 4$ **e** $7w + 6 - 2w + 2$ **h** $p^2 + p^2 + 2p + p^2 + 4p$

   c $d + 5 + 2d - 3$ **f** $2x + 3y + 4x + 2y$ **i** $4w + 2 + y - 3$

3 Simplify:

   a $a^2 + a^3 + 2a^2$ **d** $4a^2b + 5a^2b$ **g** $3cd^2 + 4cd^2 - 2dc^2 - 3c^2d$

   b $w^5 + 2w^5 + w$ **e** $5x^3y + 2xy^3 + 2x^3y$ **h** $7abc^2 + 4ab^2c + 5abc^2 + 3ba^2c$

   c $t^2 + 3t^2 + t - t^2$ **f** $6st^2 - 2st^2 + st^2$

1 Multiply out the following:

   a $2(x + 3)$ **c** $2(t + 1)$ **e** $3(d - 4)$ **g** $-2(x + 1)$

   b $4(y + 4)$ **d** $6(h - 2)$ **f** $5(3ab + 2a)$ **h** $-3(p - 2)$

2 Expand:

   a $p(p + 2)$ **c** $b(2b + 3)$ **e** $w(2w - 3)$ **g** $t(3t - 4)$

   b $y(y + 4)$ **d** $g(g - 4)$ **f** $k(3k + 4)$ **h** $a(2b - 3)$

3 Expand:

   a $p(p^2 + 4)$ **d** $w(w^2 - 3)$ **g** $x(xy - xy^2)$

   b $d(d^2 + 5)$ **e** $a^2(a^2 + 5)$ **h** $a(6ab - 2a)$

   c $t^2(t - 3)$ **f** $p(p^3 - p)$ **i** $xyz(xz + yz)$

Homework 3

1 Expand and simplify:

a $2(p+2)+3(p+4)$ **c** $7(y+2)-3(y+4)$ **e** $7(h+2)+3(2-h)$

b $4(x+4)+5(x+1)$ **d** $8(g-2)+4(g+2)$ **f** $3(2-a)-4(a-2)$

2 Find the integers a and b if $3(2m-a)+b(4m+2)=22m+5$.

3 Expand and simplify:

a $3(x+2)-2(x-2)$ **c** $2a(a+1)+4a(3+a)$

b $4p-(p+2)$ **d** $3x(y+2)+2y(x-1)$

Homework 4

1 Copy and complete the following:

a $3(p+2)=3p+...$ **e** $2(5y+1)=10y+...$ **i** $7(3t-2)=21t-...$

b $4(p+2)=4p+...$ **f** $6(h-4)=6h-...$ **j** $4(4w-5)=...-20$

c $4(k+5)=...+20$ **g** $6(g-2)=6g-...$ **k** $-3(b+1)=-3b...$

d $9(3p+4)=27p+...$ **h** $4(q-5)=...-20$ **l** $-5(y-3)=-5y...$

2 Factorise:

a $2x+10$ **c** $5p+20$ **e** $4f-12$ **g** $20y-80$

b $3p+21$ **d** $7d-21$ **f** $10a+14$ **h** $14x+21y-35z$

3 Copy and complete the following:

a $p(p+4)=p^2+...$ **c** $x(x-7)=x^2...$ **e** $4g(g-3)=...$

b $y(y+5)=...+5y$ **d** $f(f-6)=...-6f$

4 Factorise:

a $5ab+2b$ **c** $7ty+10y$ **e** $11ab-7bc$ **g** $2x^2+xy$

b $4pq+q$ **d** $3xz+5x^2z$ **f** p^2+3p **h** $2p^2q+5pq^2$

5 Match the expression with the correct factorisation. Fill in the missing expressions and factorisations.

Expression	Factorisation
d^2+5d	$2(5a-8)$
	$5(d+1)$
p^2q+pq^2	
$10a-16$	
	$d(d^2+5)$

7 Decimals

Homework 1

1 **a** Which is the tenths figure in 4.235?

    b Which is the thousandths figure?

2 Write down the place value of the:

    a 9 in 0.92   **e** 5 in 5.6746   **i** 7 in 0.0976

    b 8 in 0.78   **f** 3 in 6.2853   **j** 2 in 0.0342

    c 2 in 0.268   **g** 1 in 123.4

    d 6 in 8.2869   **h** 8 in 23.58

3 Work these out without using a calculator.

    a 3.7×10 **f** 25.14×100 **k** 9.912×1000

    b 0.342×100 **g** 0.1653×100 **l** 0.0534×100

    c 52.8×10 **h** 53.92×1000 **m** 0.1627×1000

    d 3.897×10 **i** 4.6715×1000 **n** 0.00123×100

    e 0.1259×1000 **j** $16.6267 \times 10\,000$

4 Chris says that 0.38 is bigger than 0.6, because 38 is bigger than 6.

Harry says 0.6 is bigger, because it is more than a half, and 0.38 is less than a half.

Who is right? Give a reason for your answer.

5 Find your way through this maze by following a trail of correct answers.

Start	$2.7 \div 10$ $= 2.07$	3.5×100 $= 35$	1.26×10 $= 0.126$	$3 \div 100$ $= 0.03$	End
$2.7 \div 10$ $= 0.27$	$0.2 \div 10$ $= 0.20$	0.08×10 $= 0.080$	$7 \div 100$ $= 0.07$	0.12×10 $= 1.2$	5×10 $= 5.0$
$6.4 \div 100$ $= 0.064$	1.2×1000 $= 1200$	0.1×100 $= 0.100$	$7.1 \div 10$ $= 0.71$	2.1×10 $= 20.1$	3.45×10 $= 30.45$
$2.1 \div 10$ $= 2.01$	$30 \div 1000$ $= 0.03$	$2.1 \div 10$ $= 0.021$	$7 \div 10$ $= 0.7$	1.2×100 $= 120$	0.1×100 $= 10$
$3.2 \div 100$ $= 0.32$	10×3.1 $= 31$	$1 \div 1000$ $= 0.001$	$25 \div 10$ $= 0.25$	1.2×100 $= 1200$	$3.2 \div 100$ $= 0.032$
1.3×100 $= 103$	1.02×10 $= 12$	$2 \div 1000$ $= 0.002$	$2.4 \div 100$ $= 0.024$	1.23×10 $= 12.3$	0.9×100 $= 90$

Homework 2

1 A decimal wall has bricks arranged in a triangle, like the one below.

Each brick on the bottom row has a decimal in it between 1 and 4.

For each pair of bricks, you must follow these rules:

If the number on the left is smaller than the number on the right, you add them together and write the answer in the brick above.

If the number on the left is bigger than the number on the right, you subtract.

So, as 2.4 is smaller than 3.1, you add them to get 5.5

As 3.1 is bigger than 1.2, you subtract them to get 1.9

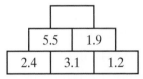

Finally, as 5.5 is bigger than 1.9, you subtract, and put the answer, 3.6, in the top brick.

Try to complete these walls using the same rules.

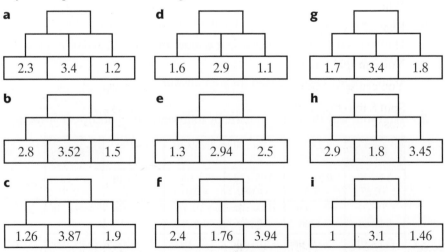

Extension: Now make up some of your own with three layers. Remember the rules, and start with numbers between 1 and 4. What is the biggest top number you can get? What is the smallest?

2 Can you find two numbers that:

    a add up to 12 and have a difference of 4

    b add up to 9 and have a difference of 4

    c add up to 4.2 and have a difference of 2

    d add up to 9.2 and have a difference of 3.8

    e add up to 0.7 and have a difference of 0.2?

Homework 3

1 Answer these:

    a 3.2×3.2 **c** 3.2×0.52 **e** 4.6×7.8 **g** 0.67×0.43

    b 4.2×1.4 **d** 4.2×9.1 **f** 5.6×3.4 **h** 0.345×0.03

2 Using your answers to question **1**, write down the answers to these questions.

    a 3.2×0.32 **c** 32×0.52 **e** 4.6×0.078 **g** 67×0.43

    b 4.2×0.14 **d** 0.42×0.91 **f** 0.0056×3.4 **h** 34.5×30

3 Neeta knows $4 \times 7 = 28$. She says that $0.4 \times 0.7 = 0.28$

    Joe says $0.4 \times 0.7 = 2.8$

    Who is right?

    Explain your answer.

4 You can split 8 into two parts, 2 and 6. Then multiply: $2 \times 6 = 12$

    You can split 8 into two parts, 3 and 5. Then multiply: $3 \times 5 = 15$

    Split 8 into two parts and multiply them together. Your aim is to get close to 14. How close can you get? (You will need to use decimals.)

5 A palindrome reads the same forwards as backwards.

So 4.2×2.4 is a palindrome. So is 1.5×5.1

a Which is bigger, 4.2×2.4 or 1.5×5.1?

b Which is bigger, 6.2×2.6 or 3.5×5.3?

c Which is bigger, 9.2×2.9 or 7.4×4.7?

In parts **a** to **c**, the two numbers have the same sum.

For example, in part **a**, $4.2 + 2.4 = 1.5 + 5.1$

d Here are two more. Can you predict which will have the bigger answer before you work them out?

  i 7.2×2.7 or 4.5×5.4

  ii 5.7×7.5 or 3.9×9.3

e Can you find any multiplications with a palindromic answer, for example, $3.3 \times 3.7 = 12.21$?

Homework 4

1 Work out $414 \div 6$?

2 Use your answer to question **1** to work these out:

  a $41.4 \div 6$ **b** $4.14 \div 0.6$ **c** $0.414 \div 0.06$ **d** $414 \div 0.6$

3 Eight friends go out for a meal. The total cost is £100.32
They decide to split the cost equally between them.

How much should they each pay?

4 Work out:

  a $2.45 \div 0.5$ **c** $2.34 \div 0.8$ **e** $9.65 \div 0.5$ **g** $8.765 \div 0.1$

  b $9.12 \div 0.6$ **d** $9.6 \div 0.12$ **f** $24.97 \div 1.1$ **h** $0.005 \div 0.4$

5 **Get Real!**

The Healthy Bite Café buys its orange juice in big containers.
Each container holds 15 litres of juice.

a How many 0.3 litre cups can they get from a 15 litre container?

b If they charge £0.85 for each cup, how much money do they get from selling all these cups?

c They pay £23.85 for a big container. How much profit do they make?

6 **Get Real!**

Aisha goes into the supermarket and sees some bottles of squash on offer at £1.10 each. She has £30 in her pocket.

a How many bottles can she buy?

b Work out the exact cost of these bottles.

c How much change will she have left?

Homework 5

Apart from question 6, this is a non-calculator exercise.

1 Change these decimals to fractions.

a 0.7   **b** 0.35   **c** 0.85   **d** 0.26   **e** 0.375   **f** 0.325

2 Change these fractions to decimals.

a $\frac{2}{5}$ **b** $\frac{3}{10}$ **c** $\frac{7}{20}$ **d** $\frac{5}{8}$

3 Change these fractions to recurring decimals.

a $\frac{2}{3}$ **b** $\frac{5}{6}$ **c** $\frac{2}{7}$ **d** $\frac{5}{12}$

4 James thinks 0.38 is the same as $\frac{3}{8}$

a Is James correct?
Give a reason for your answer.

b Which is bigger, 0.38 or $\frac{3}{8}$?

c How much bigger? Give your answer as a fraction and as a decimal.

5 **a** Change 0.5 and 0.05 to fractions.

b Now change 0.6 and 0.06 to fractions.

c What do you notice?

d If $0.65 = \frac{13}{20}$, what decimal do you think is equal to $\frac{13}{200}$?

 6 **a** Use a calculator to change these fractions to decimals.

i $\frac{1}{9}$ **ii** $\frac{12}{99}$ **iii** $\frac{123}{999}$

b There is a pattern in the questions, and a pattern in the answers.
Use the patterns to predict a fraction which is equal to
0.12345123451234512345 ...

Check you are right by changing your answer to a decimal.

8 Perimeter and area

Homework 1

1 a Find the perimeters of the following shapes made from 1 cm squares.

i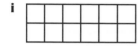

iii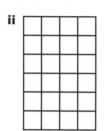

Not drawn accurately

ii

iv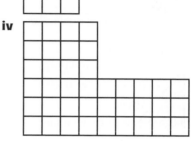

b Calculate the area of each shape.

2 Match each shape with the correct perimeter and the correct area.

a
4 cm

6 cm

b
10 cm

3 cm

c
1 cm
6 cm

Not drawn accurately

d
4 cm

i. Area = 30 cm²	**w.** Perimeter = 16 cm
ii. Area = 24 cm²	**x.** Perimeter = 14 cm
iii. Area = 16 cm²	**y.** Perimeter = 26 cm
iv. Area = 6 cm²	**z.** Perimeter = 20 cm

3 Calculate the amount of grey and white material needed to make the flag.

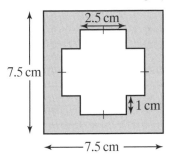

Homework 2

1 Find the area of each of these triangles.

a

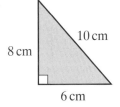

8 cm   10 cm

6 cm

c

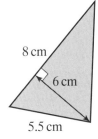

8 cm   6 cm

5.5 cm

Not drawn accurately

b

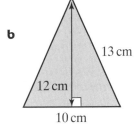

12 cm   13 cm

10 cm

d

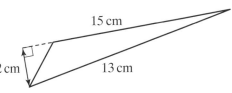

15 cm

2 cm   13 cm

2 Find the area of each of these parallelograms.

a

3 cm

8 cm

b

6.3 cm   6 cm

10 cm

c

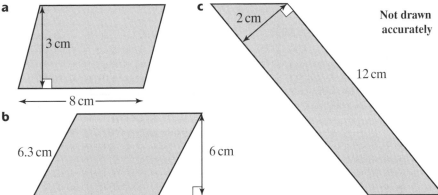

2 cm

12 cm

5 cm

Not drawn accurately

3 Fill in the gaps in the table.

	Shape (Parallelogram/Triangle)	Base	Perpendicular height	Area
a	Parallelogram	5 cm	2.5 cm	
b	Triangle	5 cm	18 cm	
c		10 cm	2.5 cm	12.5 cm²
d	Parallelogram	4 cm		12 cm²
e	Triangle	0.5 cm		8 cm²
f	Parallelogram	0.5 m	10 cm	

Homework 3

1 Estimate the area of the island where each square represents one square mile.

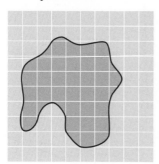

2 **a** Find the area of each of the following shapes made from 1 cm squares.

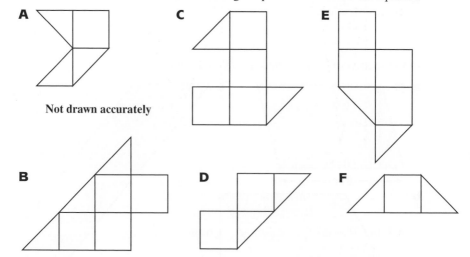

Not drawn accurately

b Draw a shape with an area of $5\frac{1}{2}$ cm².

3 Calculate the grey shaded area in each of the following:

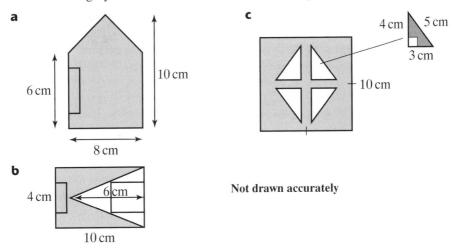

Not drawn accurately

4 Find the area of the following shapes:

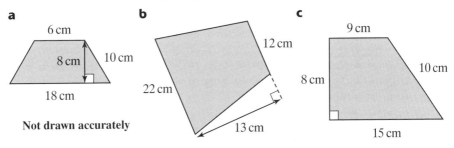

Not drawn accurately

5 Find the area of each of these kites.

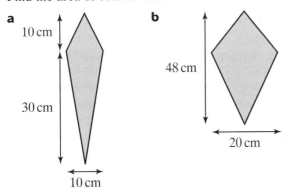

Homework 4

1 Calculate the circumference of each of the following circles:

  i leaving your answer in terms of π

   ii giving your answer to an appropriate degree of accuracy.

a
20 cm

b
8 mm

c
10 m

d
3 mm

2 Calculate the total perimeter of each of the following shapes:

  i leaving your answer in terms of π

   ii giving your answers to an appropriate degree of accuracy.

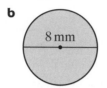

a
← 4m →

c
4 mm

b
120 cm
←2 m→

d
60°
6 cm

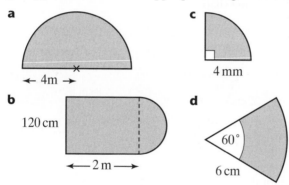

3 Complete the following table, giving your answers to an appropriate degree of accuracy.

Radius	Diameter	Circumference
	8 cm	
8 m		
		20 mm
		12π cm

4 Calculate the curved surface area of the cylinder.

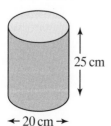

25 cm

← 20 cm →

Homework 5

1   Calculate the area of each of the following circles:

 i leaving your answer in terms of π

   ii giving your answer to an appropriate degree of accuracy.

a 7 cm   **b** 20 cm   **c** 10 m   **d** 5 mm

2   Complete the following table, giving your answers to an appropriate degree of accuracy.

Radius	Diameter	Circumference
	8 cm	
8 m		
		20 mm²
		36π cm²

3   Calculate the total area of each of the following shapes:

 i leaving your answer in terms of π

   ii giving your answer to an appropriate degree of accuracy.

a
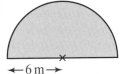
←6 m→

c

8 cm

b
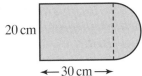
20 cm
←30 cm→

d
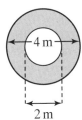
←4 m→
2 m

9 Fractions

Homework 1

1

$\frac{1}{15}$	$\frac{2}{15}$	$\frac{3}{15}$	$\frac{4}{15}$	$\frac{5}{15}$	$\frac{6}{15}$	$\frac{7}{15}$	$\frac{8}{15}$	$\frac{9}{15}$	$\frac{10}{15}$	$\frac{11}{15}$	$\frac{12}{15}$	$\frac{13}{15}$	$\frac{14}{15}$	$\frac{15}{15}$
	$\frac{1}{5}$			$\frac{2}{5}$			$\frac{3}{5}$			$\frac{4}{5}$			$\frac{5}{5}$	

Use this diagram to complete these equivalent fraction statements.

a $\dfrac{3}{5} = \dfrac{9}{\square}$ **b** $\dfrac{\square}{\square} = \dfrac{2}{5}$ **c** $\dfrac{3}{15} = \dfrac{\square}{\square}$

d Use the diagram to write down another statement about equivalent fractions.

2 Complete these equivalent fraction statements.

a $\dfrac{1}{2} = \dfrac{\square}{18}$ **b** $\dfrac{3}{9} = \dfrac{\square}{3}$ **c** $\dfrac{6}{18} = \dfrac{\square}{6} = \dfrac{\square}{3}$ **d** $\dfrac{18}{18} = \dfrac{9}{\square} = \dfrac{6}{\square} = 1$

3 Draw a diagram to show that $\frac{3}{4} = \frac{9}{12}$

4 **a** Continue this equivalent fraction pattern.
Add at least five more equivalent fractions.

$$\dfrac{1}{3} = \dfrac{2}{\square} = \dfrac{3}{\square} = \dfrac{4}{\square}$$

b What do you notice about about the number patterns in the numerators and denominators?

c Would the fraction $\frac{150}{450}$ be in the set? Explain your answer.

5 Explain why these fractions are not equivalent.

$$\frac{2}{3} \quad \frac{4}{5} \quad \frac{6}{7} \quad \frac{8}{9} \quad \frac{10}{11}$$

6 Sue says, 'If the denominator of a fraction is twice the numerator, the fraction is equivalent to one half.'

Is Sue right? Explain your answer.

7 Which fraction in this list is equal to $\frac{4}{5}$?

$$\frac{14}{15} \quad \frac{6}{8} \quad \frac{5}{6} \quad \frac{80}{100}$$

8 Show how to change $\frac{6}{8}$ into a fraction with a denominator of 56.

9 Express each of these fractions in decimal form.

   a $\frac{1}{2}$ **c** $\frac{3}{4}$ **e** $\frac{4}{5}$ **g** $\frac{67}{100}$

   b $\frac{1}{4}$ **d** $\frac{3}{10}$ **f** $\frac{1}{16}$ **h** $\frac{1}{12}$

10 Write each of these fractions in its simplest possible form.

   a $\frac{8}{10}$ **c** $\frac{5}{25}$ **e** $\frac{25}{100}$ **g** $\frac{100}{150}$

   b $\frac{40}{50}$ **d** $\frac{24}{60}$ **f** $\frac{2.4}{3.6}$ **h** $\frac{\frac{2}{3}}{\frac{4}{3}}$

11 Write down an equivalent fraction statement about hundredths, quarters and halves.

12 Get Real!

   Jane gets 16 out of 20 marks in a test. What mark out of 100 is this equivalent to?

Homework 2

1 Change these fractions to twentieths and then arrange them in order of size, smallest first.

   $\frac{7}{20}$ $\frac{2}{5}$ $\frac{3}{4}$ $\frac{7}{10}$

2 Change these fractions to hundredths and then arrange them in order of size, smallest first.

   $\frac{9}{10}$ $\frac{2}{50}$ $\frac{4}{5}$ $\frac{21}{25}$ $\frac{1}{2}$

3 Arrange these fractions in order of size, smallest first.

   $\frac{1}{2}$ $\frac{2}{5}$ $\frac{2}{50}$ $\frac{3}{4}$ $\frac{4}{5}$ $\frac{7}{20}$ $\frac{7}{10}$ $\frac{9}{10}$ $\frac{21}{25}$

4 Arrange these fractions in order of size, smallest first.

   $\frac{3}{4}$ $\frac{5}{6}$ $\frac{2}{3}$ $\frac{7}{12}$

5 **a** What common denominator would you chose for halves, thirds, and fifths?

     b Explain why it is not easy to change sevenths to tenths.

6 Kaz says, 'Hundredths are smaller than thousandths because 100 is smaller than 1000.'

   Is Kaz right? Explain your answer.

7 **a** Find a fraction between zero and one twentieth.

b Find a fraction between zero and one hundredth.

8 Get Real!

Trainers are on offer, with 20% off, in the local sports shop.

What fraction saving is this?

Homework 3

1 Work out:

a $\frac{3}{4}$ of 12   **c** $\frac{3}{4}$ of £10   **e** $150 \times \frac{3}{4}$

b $\frac{3}{4}$ of £20   **d** $\frac{3}{4} \times £18$ **f** $\frac{3}{4}$ of an hour
(give your answer in minutes).

2 Explain how you can use the answer to question **1b** to find the answer to question **1c**.

3 **a** Three quarters of a number is 12. What is the number?

b Three quarters of a number is 15. What is the number?

4 Find four fifths of:

a 100   **d** a metre (answer in centimetres)

b 50   **e** a kilogram (answer in grams).

c an hour (answer in minutes)

5 Four fifths of a number is 20. What is the number?

6 Find $\frac{2}{3}$ of:

a £15   **b** £30   **c** £18   **d** 150   **e** an hour

7 Which of these calculations will work out $\frac{3}{4}$ of 52?

a $52 \div 4 \times 3$ **c** $52 \times 4 \div 3$ **e** $52 \div 3 \times 4$

b $52 \times 3 \div 4$ **d** $52 \div 100 \times 75$ **f** $52 \times 10 \div 7.5$

8 Get Real!

A mix of concrete is made up of cement, sand and stones.

$\frac{1}{4}$ of the concrete is cement, $\frac{3}{8}$ is sand and $\frac{3}{8}$ is stones.

How much sand is there in 1000 kg of concrete?

Homework 4

1 In a class of 30 students, there are 20 girls.

  a What fraction of the class is girls?

  b What fraction of the class is boys?

2 Here are the heights of the boys in the class.

155 cm, 156 cm, 156 cm, 158 cm, 160 cm, 162 cm, 162 cm, 167 cm, 167 cm, 170 cm

  a What fraction of the boys are over 162 cm tall?

  b What fraction of the boys are less than 170 cm tall?

  c What height separates the tallest half of the boys from the shortest half?

  d The bottom $\frac{3}{10}$ of the boys are below which height?

3 Get Real!

A computer costing £900 has its price reduced by £150 in a sale.

  a By what fraction is the price reduced?

  b What fraction of the sale price is the reduction?

  c If the original price was reduced by a half what would the sale price be?

  d If the original price was increased by a half what would the new price be?

  e If the sale price of a computer was £600 after a reduction of $\frac{1}{4}$, what was the original price?

4 Ann is making nut biscuits from flour, butter, sugar and nuts. She needs 120 g of flour to make 300 g of biscuits.

  a What fraction of the biscuits is made up of flour?

  b The weight of sugar she needs is two thirds of the weight of flour. How much sugar does she need?

  c Nuts make up one fifth of the ingredients. Ann has just 45 g of butter in her fridge. How much butter will be left when she has made 300 g of biscuits?

Homework 5

You should be able to do all of these without a calculator but make sure you know how to use a calculator to check your work or to speed it up.

1 Work out:

  a $\frac{3}{10}+\frac{2}{15}$ **b** $\frac{3}{10}-\frac{2}{15}$ **c** $\frac{4}{5}+\frac{3}{8}$ **d** $\frac{4}{5}-\frac{3}{8}$

2 Work out:

a $2\frac{3}{10} + 1\frac{2}{15}$ **b** $2\frac{3}{10} - 1\frac{2}{15}$ **c** $3\frac{1}{4} + 1\frac{1}{3}$ **d** $3\frac{1}{4} - 1\frac{1}{3}$

3 Work out:

a $2\frac{2}{15} + 1\frac{3}{10}$ **b** $2\frac{2}{15} - 1\frac{3}{10}$ **c** $3\frac{1}{3} + 1\frac{1}{4}$ **d** $3\frac{1}{3} - 1\frac{1}{4}$

4 How many times does $\frac{2}{3}$ go into 4? (Work this out by repeatedly subtracting $\frac{2}{3}$ from 4.)

5 Find the sum of $1\frac{2}{3}$ and $2\frac{1}{5}$

6 **a** John says, 'When you double $2\frac{1}{2}$ you get $4\frac{1}{4}$'

     Is John right? Show how you worked out your answer.

     b Lynn says, '$\frac{4}{5}$ plus $\frac{1}{3}$ is $\frac{5}{8}$'

     Is Lynn right? Show how you worked out your answer.

7 Without working out the answers, arrange these calculations in order, the one with the smallest answer first and the one with the largest answer last.

$1 + \frac{2}{3}$ $1 - \frac{2}{3}$ $1 + \frac{4}{3}$ $1 - \frac{4}{3}$

8 Find three different pairs of fractions that add up to 6.

9 **Get Real!**

David's recipe says that he needs two thirds of a cup of milk for some ice cream for four people. How much milk does he need if he wants to make the ice cream for eight people?

10 **Get Real!**

Diane is making hats and bags for a school sale.
Each hat needs $\frac{3}{8}$ of a yard of fabric and each bag needs $\frac{2}{3}$ of a yard.
How much fabric is needed to make a hat and bag set?

Homework 6

1 Work out:

a $15 \times \frac{1}{5}$ **c** $36 \times \frac{2}{3}$ **e** $15 \times \frac{4}{5}$ **g** $36 \times \frac{5}{9}$

b $26 \times \frac{1}{2}$ **d** $45 \times \frac{2}{9}$ **f** $26 \times \frac{1}{4}$ **h** $45 \times \frac{3}{10}$

2 Work out:

a $\frac{4}{5} \times \frac{1}{2}$ **c** $\frac{3}{4} \times \frac{1}{3}$ **e** $\frac{7}{9} \times \frac{3}{7}$ **g** $\frac{2}{3} \times \frac{5}{8}$

b $\frac{1}{2} \times \frac{5}{8}$ **d** $\frac{3}{10} \times \frac{2}{15}$ **f** $\frac{4}{9} \times \frac{9}{10}$ **h** $\frac{9}{16} \times \frac{5}{9}$

3 Work out:

a $2 \div \frac{1}{7}$ **c** $12 \div \frac{1}{6}$ **e** $8 \div \frac{2}{3}$ **g** $18 \div \frac{3}{4}$

b $5 \div \frac{1}{4}$ **d** $24 \div \frac{1}{2}$ **f** $10 \div \frac{5}{8}$ **h** $32 \div \frac{4}{5}$

4 Work out:

a $\frac{5}{6} \div \frac{1}{3}$ **c** $\frac{7}{10} \div \frac{3}{5}$ **e** $\frac{2}{7} \div \frac{4}{7}$ **g** $\frac{9}{25} \div \frac{3}{5}$

b $\frac{5}{8} \div \frac{1}{4}$ **d** $\frac{2}{7} \div \frac{2}{7}$ **f** $\frac{2}{7} \div \frac{1}{7}$ **h** $\frac{4}{15} \div \frac{2}{3}$

5 Jared says that $\frac{3}{4} \div \frac{1}{2} = \frac{3}{8}$

Is he correct? Explain your answer.

6 Yasmin and Zoe are working out $\frac{9}{10} \times \frac{5}{6}$

Yasmin says the answer is $\frac{45}{60}$

Zoe says the answer is $\frac{3}{4}$

Who is correct? Explain your answer.

7 Lynn says, '6 divided by $\frac{1}{3}$ is 2.'

Is Lynn correct? Explain your answer.

8 Without working out the answers, arrange these calculations in order, the one with the smallest answer first and the one with the largest answer last.

$1 \times \frac{1}{3}$ $1 \div \frac{1}{3}$ $1 \times \frac{2}{3}$ $1 \div \frac{2}{3}$

9 **Get Real!**

David's recipe says that he needs two thirds of a cup of cream for some ice cream for four people. How much cream does he need if he wants to make ice cream for six people?

10 **Get Real!**

Diane has 4 yards of fabric to make some hats for a school sale. Each hat needs $\frac{3}{8}$ yard of fabric. How many hats can Diane make and how much fabric is left over?

1 The table shows the sales of teas.

Tea	Ordinary	Apple	Blackcurrant	Lemon	Other
Frequency	18	8	4	5	1

Show this information as:

a a pictogram

b a bar chart

c a pie chart.

2 The table shows how 90 students travelled to college.

Travel	Car	Bus	Taxi	Walk	Cycle
Frequency	43	9	5	20	13

Show this information as:

a a bar chart

b a pie chart.

3 The graph shows a dual bar chart for the number of phone calls and texts received on five days during the week.

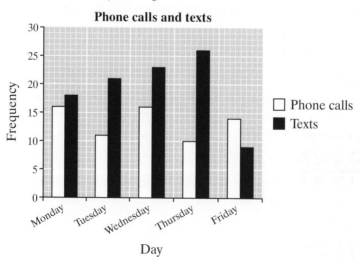

a Copy and complete the table for the number of phone calls and texts received.

Day	Phone calls	Texts
Monday		
Tuesday		
Wednesday		
Thursday		
Friday		

b What is the modal number of phone calls?

c Calculate the range for the number of texts.

d Which day had the greatest number of phone calls and texts altogether?

4 The pictogram shows the number of books borrowed from a library.

Day	Number of books borrowed
Monday	
Tuesday	
Wednesday	
Thursday	
Friday	

Key = 2 books

a How many books were borrowed on Tuesday?

b How many books were borrowed on Friday?

c On what day were the most books borrowed?

d What is the mean of the number of books borrowed over the five days?

e How many books were borrowed on Thursday?

f Give a possible reason for your answer in part **e**.

5 The table shows the results of a survey to find students' favourite pets.

Pet	Tally
Cat	JHT JHT JHT JHT JHT
Dog	JHT JHT JHT JHT JHT II
Horse	JHT JHT IIII
Bird	JHT JHT JHT I
Fish	JHT III

Draw a pie chart to show the information.

6 The table shows the minimum and maximum temperatures at a seaside resort.

Day	Minimum temperature (°C)	Maximum temperature (°C)
Monday	14	19
Tuesday	11	18
Wednesday	13	22
Thursday	14	23
Friday	17	25

Show this information in a dual bar chart. Use your graph to make five different comments about the minimum and maximum temperature.

7 Students at a college are asked to choose their favourite colour. Their choices are shown in the pie chart below:

Favourite film

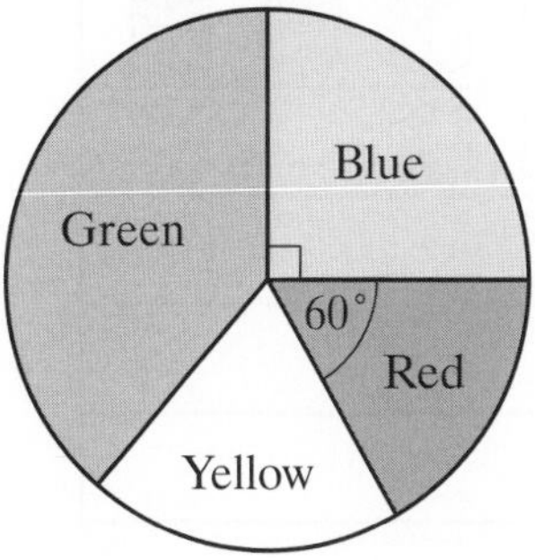

A total of 45 students chose the colour blue.

a How many students were included in the survey?

b How many students chose red?

Twice as many students chose green as chose yellow.

c How many students chose green?

8 Write down one advantage and one disadvantage of using each of the following representations:

a pictogram

b bar graph

c pie chart.

Homework 2

1   Janesh undertakes a survey on packets of chocolate beans. He counts the number in each packet. The results are as follows:

30	31	30	30	29	32	41	29	30	31
30	28	30	31	30	30	30	30	31	30

   a Show this information in a stem-and-leaf diagram.

   b What is the most likely number of chocolate beans in a packet?

   c Which number in the list is likely to be an error?

2   The heights, in centimetres, of 20 boys were recorded as follows.

130	128	120	150	140
173	135	149	160	125
135	155	143	170	165
168	128	151	132	149

Construct a stem-and-leaf diagram to represent this information.

Heights of 20 boys

Stem	Leaf
12	
13	
...	

Key: 12|5 represents 125 cm

3   The prices paid for sandwiches in a canteen are shown below:

£3.25   £2.65   £3.15   £1.95   £2.85   £2.65   £3.25   £3.75   £1.95   £2.95

Copy and complete the following stem-and-leaf diagram to show this information.

Prices paid for sandwiches

Stem (£)	Leaf (pence)
1	
2	
3	

Key: 3|25 represents £3.25

4 The marks obtained in a test were recorded as follows.

11   22   20   19   25   8   12   9   27   22   10   18   24   16   15   22

  a Show this information in an ordered stem-and-leaf diagram.

  b What was the least number of marks in the test?

  c Write down the mode of the marks in the test.

  d Write down the range of the marks in the test.

5 The heights of 15 students were recorded as follows.

Heights of students

Stem (feet)	Leaf (inches)
4	09 10
5	00 07 08 11 07 10 08 03 08 01 06
6	02 01

Key: 5|06 represents 5 feet 6 inches

  a Show this information in an ordered stem-and-leaf diagram.

  b What was the tallest height recorded?

  c Write down the range of the heights.

6 The marks scored in a test by 15 boys and 15 girls is shown in this back-to-back stem-and-leaf diagram.

Number of marks in a test

Leaf (units) Girls	Stem (tens)	Leaf (units) Boys
9	0	6 7 8 9
9 8 6 5 4 2 0	1	1 3 4 5 5 5 7 8 9
7 6 4 3 2 1	2	2 4
0	3	

Key: 3|2 represents 23 marks   **Key:** 2|3 represents 23 marks

  a Calculate the median for the girls.

  b Calculate the mode for the boys.

  c Calculate the mean for the girls.

  d Calculate the range for the boys.

  e What can you say about the performance of girls and boys?

7 Andy measures the resting heartbeat of 15 children and 15 adults. His results are recorded in the table below.

	Resting heartbeat														
Child	69	71	73	76	79	80	81	83	83	83	85	87	88	89	94
Adult	65	67	69	69	70	73	74	75	75	75	78	79	80	81	84

a Show this information in a back-to-back stem-and-leaf diagram.

b Andy says that 'the older you are the slower your resting heartbeat.' Use your data to check Andy's hypothesis.

Homework 3

1 The table shows the time spent waiting at a clinic.

Time, t (minutes)	Frequency
$0 \leqslant t < 10$	4
$10 \leqslant t < 20$	14
$20 \leqslant t < 30$	4
$30 \leqslant t < 40$	1

Draw a frequency diagram to represent the data.

2 The following table shows the temperature of a patient at different times of the day.

Time of day	10:00	11:00	12:00	13:00	14:00	15:00
Temperature (°F)	102.5	101.3	102	100.6	100.1	99.6

a Draw a line graph to show the temperature.

The normal temperature is 98.8 °F.

b At what time is the patient's temperature likely to return to normal?

3 The frequency diagram shows the ages of people in a church.

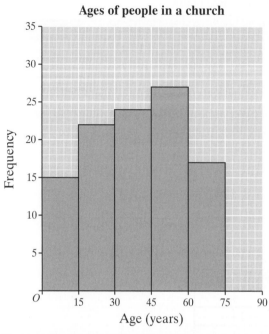

Copy and complete this table to show this information.

Age, y (years)	Frequency
$0 \leqslant y < 15$	
$15 \leqslant y < 30$	
$30 \leqslant y < 45$	
$45 \leqslant y < 60$	
$60 \leqslant y < 75$	

4 The table shows the minimum and maximum temperatures at a seaside resort.

Day	Minimum temperature (°C)	Maximum temperature (°C)
Monday	15	19
Tuesday	17	22
Wednesday	14	21
Thursday	13	19
Friday	11	16

Draw a line graph to show:

a the minimum temperature

b the maximum temperature.

Use your graph to find:

c the day on which the lowest temperature was recorded

d the day on which the highest temperature was recorded.

5 Narinder's termly exam results are shown in the following table.

Year	2004	2005	2005	2005	2006	2006	2006
Session	Autumn	Spring	Summer	Autumn	Spring	Summer	Autumn
Results (%)	98	93	70	97	95	69	98

a Show this information on a graph.

b Narinder says that her performance is improving. Is she correct? Give a reason for your answer.

6 The following table shows the mean number of sales of CDs and DVDs at a corner shop from 1998 to 2004.

Year	1998	1999	2000	2001	2002	2003	2004
CD sales	320	345	310	305	300	235	260
DVD sales	100	125	165	220	230	240	280

a Show this information on a graph.

b Comment on your results.

11 Scatter graphs

1 Draw the scales and plot the points given.

a

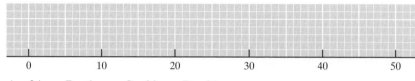

| 0 | 10 | 20 | 30 | 40 | 50 |

A = 24   B = 6   C = 38   D = 21
E = 7   F = 41   G = 12   H = 29

b

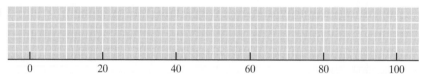

| 0 | 20 | 40 | 60 | 80 | 100 |

A = 6   B = 24   C = 94   D = 44   E = 34
F = 78   G = 52   H = 70   I = 84   J = 60

c

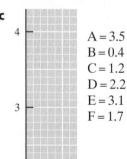

A = 3.5
B = 0.4
C = 1.2
D = 2.2
E = 3.1
F = 1.7

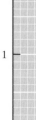

2 Write down the value of each of the points given below.

a

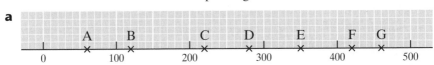

b

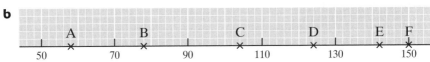

c

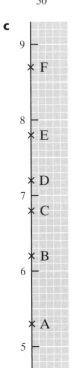

3 The table shows the ages of ten family members and the number of holidays taken abroad. Copy the axes and plot the points given in the table.

Age (years)	42	8	36	47	14	22	31	15	40	31
Number of holidays abroad	16	1	21	15	3	4	9	5	22	11

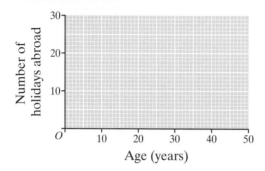

4 The table shows the number of matches played and the number of goals scored. Copy the axes and plot the points given in the table.

Matches	2	12	8	6	7	11	3	15
Goals scored	35	7	25	20	10	14	26	2

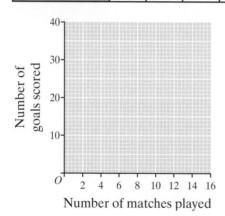

5 The table shows the ages of contestants and the number of medals won.
Copy the axes and plot the points given in the table.

Age (years)	16	18	28	22	23	19	27
Medals won	20	14	30	64	46	34	74

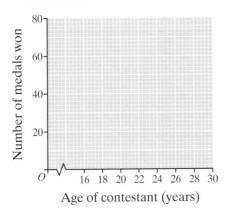

6 The graph shows the marks awarded by two judges in a competition.

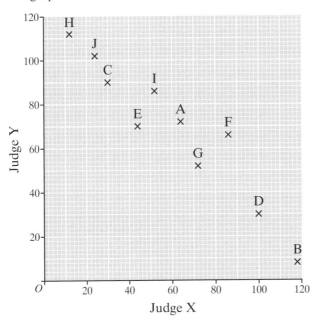

a Copy and complete the table for the marks awarded by the two judges.

	A	B	C	D	E	F	G	H	I	J
Judge X										
Judge Y										

b Are the two judges consistent? Give a reason for your answer.

Homework 2

1   For each of the following scatter graphs, match the diagram to the description.

a

b

c

d

i Strong positive correlation

ii Strong negative correlation

iii No correlation

iv Weak positive correlation

v Weak negative correlation

2 For each of the following data sets:

   i describe the type and strength of correlation

   ii write a sentence explaining the relationship between the two sets.

  a The age of a person and their IQ.

  b The number of cars on the road and the time taken to get home.

  c The age and value of a computer.

  d The distance to the holiday destination and the cost of a holiday.

3 The scatter graph shows a comparison of the prices of paintings in 2000 and 2006.

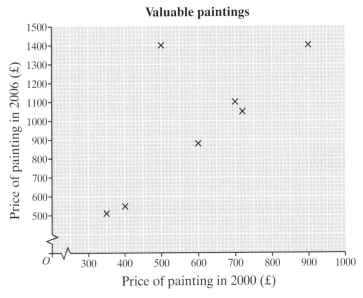

Valuable paintings

  a Describe the type and strength of correlation.

  b Write a sentence explaining the relationship between the two sets of data.

  c Does the relation hold for all of the paintings? Give a reason for your answer.

4 The table shows the ages and second-hand values of eight cars.

Age of car (years)	3	1	4	7	12	9	8	2
Value of car (£)	2900	4000	2100	1200	300	500	1400	3500

  a Draw a scatter graph of the results.

  b Describe the type and strength of correlation.

  c Write a sentence explaining the relationship between the two sets of data.

5 The table shows the rainfall and the number of raincoats sold at a department store.

Amount of rainfall (mm)	0	1	2	5	6	9	11
Number of raincoats sold	12	25	48	56	63	85	98

a Draw a scatter graph of the results.

b Describe the type and strength of correlation.

c Write a sentence explaining the relationship between the two sets of data.

6 The table shows the history and physics results for nine students.

History (%)	10	20	76	74	40	62	70	26	19
Physics (%)	75	82	15	23	51	33	18	64	78

a Draw a scatter graph of the results.

b Describe the type and strength of correlation.

c Write a sentence explaining the relationship between the two sets of data.

Homework 3

1 The table shows the ages and second-hand values of eight cars.

Age of car (years)	3	1	4	7	11	9	8	2
Value of car (£)	2900	4000	2100	1200	300	500	1400	3500

a Draw a line of best fit on the scatter graph you drew in Homework **2**, question **4**.

b Use your line of best fit to estimate:

i the value of a car if it is 10 years old

ii the age of a car if its value is £2000.

2 The table shows the rainfall and the number of raincoats sold at a department store.

Amount of rainfall (mm)	0	1	2	5	6	9	11
Number of raincoats sold	12	25	48	56	63	85	98

a Draw a line of best fit on the scatter graph you drew in Homework **2**, question **5**.

b Use your line of best fit to estimate:

i the number of raincoats sold for 8 mm of rainfall

ii the amount of rainfall if 50 raincoats are sold.

3 The table shows the history and physics results for nine students.

History (%)	10	20	76	74	40	62	70	26	19
Physics (%)	75	82	15	23	51	33	18	64	78

a Draw a line of best fit on the scatter graph you drew in Homework **2**, question **6**.

b Use your line of best fit to estimate:

   i the physics result for a student who scores 54 in history

   ii the history result for a student who scores 90 in physics.

c Which of these two answers is likely to be a better estimate? Give a reason for your answer.

4 The scatter graph shows a line of best fit for the science and maths results of eight students.

Science and maths results

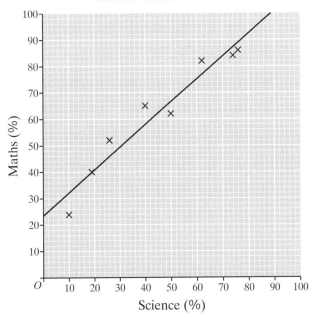

a Use the line of best fit to estimate:

   i the maths result for a student who scores 60 in science

   ii the science result for a student who scores 100 in maths.

b Which of these two answers is likely to be a better estimate? Give a reason for your answer.

5 The table shows the engine sizes and maximum speeds of eight cars.

Engine size (cc)	Maximum speed (mph)
1100	80
1800	125
2900	142
1400	107
1300	96
1000	85
2500	135
2000	131

a Draw a scatter graph and the line of best fit of the results.

b Describe the relationship between a car's engine size and its maximum speed.

c Use your line of best fit to estimate:

  i the maximum speed of a car with an engine size of 1500 cc

  ii the engine size of a car whose maximum speed is 150 mph.

d Explain why your last answer might not be accurate.

6 A line of best fit shows the relationship between two sets of data in an experiment.

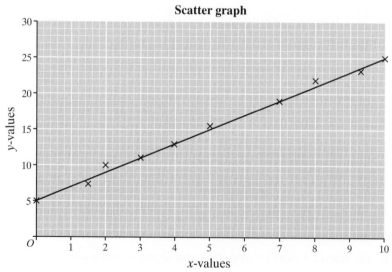

Scatter graph

Use the line of best fit to estimate:

a the value of y when $x = 7$

b the value of x when $y = 22$.

7 The following table gives the age and speed of 15 people over a distance of 100 m.

Age (years)	12	5	8	10	11	12	15	14	18	28
Speed (m/s)	5.4	1.8	3.6	4.2	5.8	6.1	6.8	7.4	8.1	7.3

a Draw a scatter graph of the results.

b Draw a line of best fit for the results.

c Describe the relationship between speed and age.

d Find an estimate for the speed of someone aged 13 years.

12 Properties of polygons

Homework 1

1 Calculate the angles marked with letters in these diagrams.

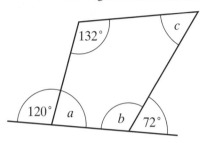 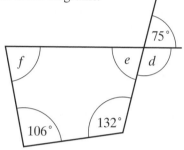

2 A quadrilateral has three angles of 86°, 100° and 81°. Calculate the size of the fourth angle.

3 A parallelogram has an exterior angle of 74°. Draw a sketch of the parallelogram, and work out the size of all four interior angles.

4 Chris draws a quadrilateral with three angles of 80° each.

   a Calculate the fourth angle.

   b Name the shape he has drawn.

5 This tessellation is made up of three different shapes.

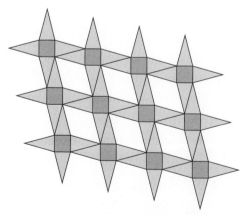

   a What is the name of each shape?

   b The obtuse angles in the diagram are all 120°.
   What are the angles of each shape?

6 Is it possible to draw:

a a quadrilateral with three obtuse angles?

b a quadrilateral with a reflex angle and an obtuse angle?

c a quadrilateral with a reflex angle and two obtuse angles?

d a trapezium with two right angles?

e a trapezium with exactly one right angle?

7 Arthur, Ben and Charlotte measure the angles of the same quadrilateral. Here are their results.

Arthur: 67°, 112°, 78°, 113°

Ben: 112°, 68°, 68°, 112°

Charlotte: 102°, 78°, 68°, 112°

a Who must be wrong?

b If the shape is a rhombus, who has measured correctly?

Homework 2

1 Follow the flowchart for each of these shapes.

SQUARE RHOMBUS KITE TRAPEZIUM
PARALLELOGRAM RECTANGLE

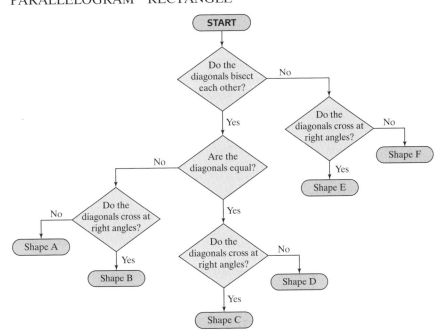

Write your answers as: Shape A is a ...

2 Calculate the angles marked with letters in the diagrams below.
You will need to use the properties of diagonals.
Explain how you know the size of the angles.

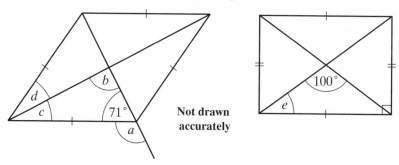

Not drawn accurately

3 A trapezium has an angle of 78° and another of 123°.
What are the other two angles?

4 The diagram shows two congruent parallelograms and a square.

Calculate the sizes of the angles in the parallelogram.

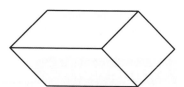

5 a Draw two lines, both 6 cm long. They must cross each other, exactly in the middle of each line, so there is 3 cm on each side of the crossing point:

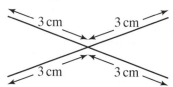

Join the ends to make a quadrilateral. Name the shape you have drawn.

b Draw two lines, one 6 cm long and the other 10 cm long. They must cross each other, exactly in the middle of each line.

Join the ends to make a quadrilateral. Name the shape you have drawn.

c Draw two lines, one 6 cm long and the other 10 cm long. They must cross each other, exactly in the middle of each line. They must cross at right angles.

Join the ends to make a quadrilateral. Name the shape you have drawn.

d Draw two lines, both 6 cm long. They must cross each other, exactly in the middle of each line. They must cross at right angles.

Join the ends to make a quadrilateral. Name the shape you have drawn.

e Draw two lines, one 6 cm long and the other 10 cm long. They must cross at right angles. They must cross each other, exactly in the middle of the 6 cm line, but not in the middle of the 10 cm line.

Join the ends to make a quadrilateral. Name the shape you have drawn.

Homework 3

1 Calculate the angles marked with letters in the diagram.
Give a reason for your answers.

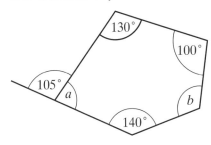

Not drawn accurately

2 A hexagon has three angles of 130° and two of 112°.
Calculate the sixth angle.

3 A regular polygon has an exterior angle of 20°.
How many sides does it have?

4 Draw a circle with a radius of 6 cm. Draw a regular decagon that just fits
inside the circle.

5 The diagram below shows a regular octagon ABCDEFGH and a regular
hexagon DCIJKL.

Calculate the angles:

a HAB   **c** BCI   **e** LKC

b CIJ   **d** BCG   **f** GCK

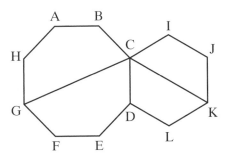

6 Explain why a regular hexagon will
tessellate, but a regular pentagon
will not.

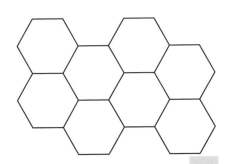

13 Indices

Homework 1

Apart from question 3, this is a non-calculator exercise.

1

16	−8	$\frac{1}{3}$
−11	0.4	$\frac{4}{9}$
$\frac{1}{4}$	−1	14

From the numbers in the grid above write down:

a a square number
b a cube number
c the square root of 196
d the cube root of −1
e $\left(\frac{1}{2}\right)^2$

2 Write down the value of each of these.

a 6^2
b 13^2
c 2.5^2
d 5^3
e 100^3
f $\sqrt[3]{27\,000}$
g $\sqrt{121}$
h $\sqrt{10\,000}$
i $\sqrt[3]{-216}$
j $\sqrt{36}$
k $\sqrt[3]{-27}$

3 Write down the value of each of these.

a 3.3^2
b 9.9^2
c 1.001^2
d 2.01^3
e 1000.1^3
f $\sqrt{1000}$
g $\sqrt{0.5}$
h $\sqrt[3]{9}$
i $\sqrt[3]{-0.001}$
j $\sqrt[3]{0.001}$

4 David says that the square of a number is always bigger than the number. Is he correct?

Give a reason for your answer.

5 Get Real!
If the area of a square is 289 mm², calculate the length of the square.

289 mm²

Homework 2

 1 Write these in index notation.

   a $6 \times 6 \times 6 \times 6 \times 6$ **c** $17 \times 17 \times 17$

   b $2 \times 2 \times 2 \times 2 \times 2 \times 2 \times 2 \times 2 \times 2 \times 2$ **d** 10

 2 Find the value of each of the following.

   a 9^2 **c** 3^3 **e** 1^3 **g** $\dfrac{2}{3^2}$

   b 2^8 **d** 10^1 **f** $\dfrac{1}{3^2}$

3 Find the value of each of the following.

   a 2^{20} **d** 8^3 **g** $2^{10} - 2^9$

   b 2^{21} **e** 5^{10} **h** $10^{51} \div 10^{49}$

   c 3^8 **f** -2^6

4 Jenny says that $10^{10} - 10^9 = 10^{10-9} = 10^1 = 10$. Is she correct?

   Give a reason for your answer.

 5 Simplify the following.

   a $3^8 \times 3^{11}$ **c** $\dfrac{3^7}{3^6}$ **e** $(3.1^2)^{10}$

   b $\dfrac{12^4}{12^2}$ **d** $1.5^{10} \div 1.5^9$

6 Simplify the following.

   a $a^5 \times a^3$ **c** $\dfrac{c^{11}}{c}$ **e** $q^{10} \div q^{11}$

   b $b^{10} \times b^6$ **d** $p^{100} \div p^{99}$ **f** $(s^2)^4$

7 The number 256 can be written as 16^2 in index form.

   Write down three other ways that 256 can be written in index form.

8 Tariq is investigating square numbers. He says that square numbers cannot end in a 2. Is he correct? Give a reason for your answer.

Homework 1

1 Draw the next two diagrams in each of the following sequences.

a

b

c

d

2 Write the next three terms in each of the following sequences.

    a 3, 6, 9, 12, ...   **c** 2, 12, 22, 32, ...   **e** 2, 4, 8, 16, ...

    b 7, 11, 15, 19, ...   **d** −5, 0, 5, 10, ...   **f** 1, 10, 100, 1000, ...

3 Write the 10th and 11th numbers in the following sequences.

    a 1, 3, 5, 7, ...   **c** 11, 22, 33, 44, ...   **e** 0.1, 0.2, 0.3, ...

    b 4, 8, 12, 16, ...   **d** 1001, 1002, 1003, 1004, ...

4 Copy and complete the following table.

Pattern (*n*)	Diagram	Number of matchsticks (*m*)
1		4 matchsticks
2		7 matchsticks
3		10 matchsticks
4		
5		

a What do you notice about the pattern of matches above?

b How many matches will there be in the 10th pattern?
Give a reason for your answer.

5 Fill in the missing numbers in the following sequences.

a 5, 7, 9, ..., 13, 15

c 36, 30, 24 ..., 12

b 7, 14, 21, ..., 35, 42

d − 10, ..., − 2, 2, 6

6 Write down the term-to-term rule for each of the following sequences.

a 7, 11, 15, 19 ...

d 3, 5.5, 8, 10.5, ...

g 21, 17, 13, 9, ...

b 0, 9, 18, 27, ...

e 2.1, 3.2, 4.3, 5.4, ...

h 8, 4, 2, 1, $\frac{1}{2}$, ...

c 2, 4, 8, 16, 32, ...

f 0.1, 1, 10, 100, ...

7 Write down the term-to-term rule for the following diagrams.

What is the special name given to the sequence of numbers from these diagrams?

8 Here is a sequence of numbers.

3   5   9   17   33

The rule for continuing this sequence is: **multiply by 2 and subtract 1**

a What are the next **three** numbers in this sequence?

The same rule is used for a sequence that starts with the number − 4.

b What are the next **three** numbers in this sequence?

Homework 2

1 Write the first five terms of the sequence whose nth term is:

    a $n+7$ **c** $5n-4$ **e** $2n^2$ **g** n^2+2

    b $3n$ **d** n^2-5 **f** $3n-7$ **h** $\dfrac{n+1}{n+2}$

2 Write down the nth term in each of the following linear sequences:

    a $1, 2, 3, 4, \ldots$ **d** $33, 31, 29, 27, \ldots$ **g** $1.5, 4, 6.5, 9, \ldots$

    b $0, 4, 8, 12, \ldots$ **e** $1000, 995, 990, \ldots$ **h** $-13, -5, 3, 11, \ldots$

    c $7, 13, 19, 25, \ldots$ **f** $-4, -1, 2, 5, \ldots$

3 Write down the formula for the number of squares in the nth pattern.

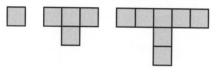

4 Look at the following table.

Pattern (n)	Diagram	Number of matchsticks (m)
1		3 matchsticks
2		5 matchsticks
3		7 matchsticks

Write down the formula for the number of matches (m) in the nth pattern.

5 Write the nth term in each of the following non-linear sequences:

    a $1, 4, 9, 16, \ldots$ **d** $1, 8, 27, 64, 125, \ldots$

    b $0, 3, 8, 15, \ldots$ **e** $2, 9, 28, 65, 126, \ldots$

    c $11, 14, 19, 26, \ldots$ **f** $1, 10, 100, 1000, \ldots$

6 Write the nth term in each of the following sequences:

    a $1 \times 2 \times 3, 2 \times 3 \times 4, 3 \times 4 \times 5, \ldots$ **c** $\dfrac{2}{4}, \dfrac{3}{5}, \dfrac{4}{6}, \dfrac{5}{7}, \ldots$

    b $\dfrac{1}{2}, \dfrac{1}{3}, \dfrac{1}{4}, \dfrac{1}{5}, \ldots$ **d** $0.01, 0.02, 0.03, 0.04, \ldots$

15 Coordinates

Homework 1

1 Write down the coordinates of:
   a the head of the zebra
   b the tail of the crocodile
   c the top of the oil rig.

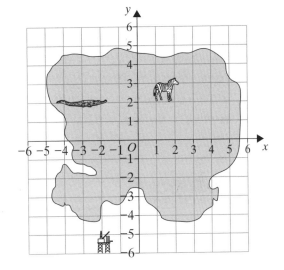

2 Draw a grid with each axis going from −5 to 5.
   a On your grid, mark the points A(5, 2), B(1, −2), and C(−3, 2).
   b Join A to B, B to C, and C back to A.
   c What is the name of the shape you have drawn?

3 Sam says that (4, 2), (2, 4), (−4, −2) and (−2, −4) are the vertices
   of a rectangle.
   Is he right?

4 (2, 5), (−1, 4) and (3, 2) are three vertices of a square.
   Where is the fourth vertex?

5 Which of these sets of points are the vertices of a square?
   a (10, 3), (9, 8), (4, 7), (5, 2)
   b (4, 4), (8, 5), (9, 9), (5, 8)
   c (−4, −1), (−2, 2), (1, 0), (−1, −3)
   d (9, 27), (26, 26), (25, 9), (8, 10)

6 (1, 1) and (−3, 1) are two vertices of a square.

Write down the coordinates of two points which could be the other two vertices.

(This question has three possible answers. Can you find all six pairs of coordinates?)

7 Gus, Josh and Asma all draw a grid, and mark the points (2, 3), (4, −1) and (0, 0). Their teacher asks them to mark a fourth point so that they have marked the corners of a parallelogram.

Gus marks (6, 2).

Josh marks (−2, 4).

Asma marks (2, −4).

a Who is correct?

b What do you notice about the six points they have marked?

Homework 2

1 **a** Draw a grid with the x-axis and y-axis going from −6 to 6.

   b Mark four points with a y-coordinate of 4.

   Draw a straight line through all four points.

   c What is the equation of the line you have drawn?

   d Now draw four points with an x-coordinate of −5.

   Draw a straight line through all four points.

   e What is the equation of the line you have drawn?

   f Where do the two straight lines cross?

   g Now draw the line $x = -2$.

   h What other line do you need to draw to complete a square?

2 Write down the coordinates of the points where these lines cross.

   a $x = 3$ and $y = 1$ **d** $x = 0$ and $y = -3$

   b $x = -5$ and $y = 2$ **e** $x = -4.5$ and $y = -2.5$

   c $x = -3$ and $y = -7$ **f** $y = -3$ and $y = x$

3 Look at this rule: The *y*-coordinate must be three more than the *x*-coordinate.

These two points fit the rule: (1, 4) and (−2, 1).

Write down three more pairs of coordinates that fit the rule.

Plot the points on a coordinate grid.

What is the equation of the line that passes through all the points?

4 Kazu says that the point (3, 6) is on the line $y = x + 3$.

Bob says that the point (3, 6) is on the line $y - x = 3$.

Who is right? Give a reason for your answer.

5 For each grid below:

  i write down the coordinates of three points on the line

  ii use your answers to part **i** to help you write down the equation of the line.

a
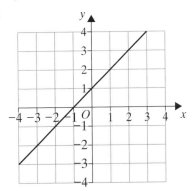

c
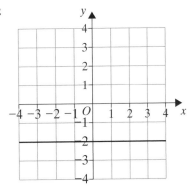

b
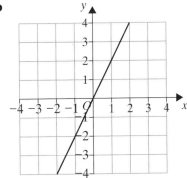

<u>6</u> For each grid below:

   i write down the coordinates of three points on the line

   ii use your answers to part **i** to help you write down the equation of the line.

a

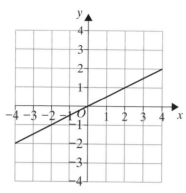

c

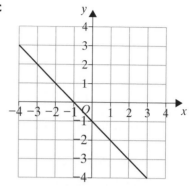

b

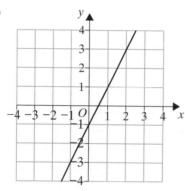

Homework 3

1 Work out the coordinates of the point halfway between (2, 10) and (4, 2).

2 Work out the coordinates of the point halfway between (2, −7) and (5, 3).

3 If A is the point (2, −4) and B is the point (−3, −3), what are the coordinates of the midpoint of the line AB?

4 Jon says that the point (−2.5, 3) is halfway between (−4, 9) and (1, 15). Is he correct? Give a reason for your answer.

5 X is the midpoint of the line AB. The coordinates of X are $(-3, 3)$.
B is the point $(1, -2)$. What are the coordinates of A?

6 Find your way through the maze by only occupying squares where M is the midpoint of AB.

1	2	3	4	5	6
START	A (5, 2) B (3, 4) M (4, 3)	A (−2, 1) B (3, −4) M (0.5, −1.5)	A (4, 7) B (−2, −3) M (1, 2)	A (−1, 9) B (−5, 2) M (−3, 4.5)	A (3, 2) B (2, −3) M (2.5, 0.5)
7 A (4, 1) B (2, −2) M (3, 0.5)	8 A (−1, 2) B (2, −3) M (0.5, 0.5)	9 A (1, 3) B (2, −2.5) M (1.5, 0.5)	10 A (−1.7, 3.2) B (1.1, −2.6) M (−0.3, 0.3)	11 A (2, −0.8) B (−1, 0.9) M (0.5, 0.05)	12 A (6, 1) B (−6, −2) M (0, −0.5)
13 A (5, −4) B (−3, 4) M (1, 0)	14 A (−3, −2) B (2, −6) M (−0.5, −4)	15 A (−1.3 2.4) B (−1.2, 9) M (−1.25, 5.7)	16 A (2, 7) B (−3, 4) M (−0.5, 6.5)	17 A (3, −2) B (3, 2) M (6, 0)	18 A (2, 1) B (−12, 5) M (−5, 3)
19 A (7, 0) B (0, 3) M (3.5, 1.5)	20 A (−1, −2.4) B (−2, 4.2) M (−1.5, 1.1)	21 A (7, −1) B (−4, −8) M (1.5, −4.5)	22 A (−8, 4) B (4, 2) M (−2, 3)	23 A (3, −3) B (1, −2) M (2, −2.5)	24 A (1.2, 2.7) B (2.4, 1.3) M (1.8, 2)
25 A (−2, −2) B (1, 5.8) M (−0.5, 1.9)	26 A (−3, −2.4) B (2, 6) M (−0.5, 1.8)	27 A (1, 2.3) B (−2, −6.2) M (0.5, −1.95)	28 A (0, 4.2) B (−2, 3.2) M (1, 3.7)	29 A (−1, 2.5) B (3, −0.1) M (2, 1.2)	30 A (2.6, 1.7) B (−1.4, 1.1) M (−0.6, 1.4)
31 A (91, 36) B (19, 8) M (50, 22)	32 A (5, −2) B (−3, 3) M (1, 0.5)	33 A (1, 3) B (−6, −3) M (−2.5, 0)	34 A (−2.4, 1.4) B (−2.2, −0.3) M (−2.3, 0.55)	35 A (0.2, −3) B (−2.3, 1) M (−1.05, −1)	36 **END**

7 **a** C is the midpoint of AB, where A is the point $(4, 1)$ and B is $(-2, 7)$.
Find the coordinates of C.

b What is the equation of the line AB?

c C is also the midpoint of the line EF, where E is the point $(2, 3)$. Find the coordinates of F.

Homework 4

1 **a** In the diagram, A is the point $(0, 4, 2)$. D is the point $(5, 0, 0)$.

Write down the coordinates of points B and C.

b What are the coordinates of the midpoints of

i OD

ii CD

iii AB

iv OB?

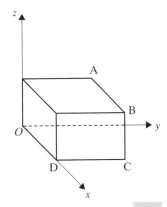

2 The diagram shows a triangular prism.

The triangular ends are isosceles.

The height of the prism is 8 units.

The base of the triangle AB is 5 units.

The length of the prism OA is 6 units.

Write down the coordinates of A, B, C and D.

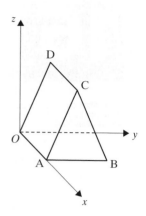

3 A cuboid has three of its vertices at (4, 1, 0), (4, 1, 6) and (2, 5, 0).

Find the coordinates of the other five vertices. As always, drawing a diagram will make it much easier.

4 The diagram shows three cubes.

The bottom cube has sides of length 8 cm, and has a vertex at (0, 0, 0).

The centre cube has sides of length 6 cm, and sits exactly in the middle of the bottom cube.

The top cube sits exactly in the middle of the centre cube, and has sides of length 4 cm.

What are the coordinates of the four top vertices of the top cube?

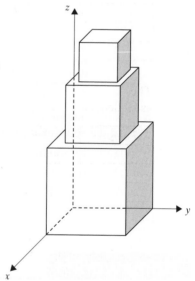

16 Collecting data

1   Use the following information to complete the tally chart.

3	1	2	4	3	2	0	2	3	3	2	0
4	2	2	3	1	1	0	3	2	1	1	3
2	4	2	1	0	0	3	4	3	2	2	

Number	Tally	Frequency
0		
1		
2		
3		
4		

2   The tally chart shows the number of people living in the same household as students in a college.

People	Tally	Frequency
1	ЖН IIII	
2	ЖН ЖН II	
3	ЖН	
4	II	
5	I	

a Copy and complete the frequency column.

b Use your table to answer the following.

   i   How many students were surveyed altogether?

   ii How many students had three people living in the same household?

   iii What is the modal number of people?

3 The following information shows the marks obtained in a test.

16	18	19	23	24	25	22	15	16	11	10
19	22	25	25	18	10	15	16	19	17	25
23	20	18	12	16	18					

Use this information to complete the tally chart below.

Number	Tally	Frequency
10–13		
14–17		
18–21		
22–25		

4 The tally chart shows the arm spans of students in a class.

Arm span (cm)	Tally	Frequency
$150 < h \leqslant 155$	III	
$155 < h \leqslant 160$	JHT IIII	
$160 < h \leqslant 165$	JHT JHT III	
$165 < h \leqslant 170$	JHT JHT JHT II	
$170 < h \leqslant 175$	JHT III	

a Copy and complete the table and use it to answer the following:

i How many students were surveyed altogether?

ii How many students had arm spans between 160 and 165 cm?

iii How many students had arm spans shorter than 160 cm?

b What is the modal class for the distribution?

5 Forty adults were asked if they were left-handed or right-handed.

	Men	Women
Left-handed	5	8
Right-handed	19	8

Use the table to answer the following questions.

a How many men were in the survey?

b How many people are right-handed?

c What fraction of those asked were right-handed men?

6 The following survey was undertaken at a dance school.

	Wears shoes	Does not wear shoes
Male	6	8
Female	5	9

a How many people were surveyed altogether?

b How many males were surveyed?

c How many females were surveyed?

d How many people wear shoes?

e How many females do not wear shoes?

7 Billy undertakes a survey of students at his college.

He asks them about their computer connections and mobile phone.

	Mobile phone with camera	Mobile phone without camera
Computer without internet	62	15
Computer with internet	8	24

a How many people did Billy ask altogether?

b How many people have a mobile phone without a camera?

c How many students have a computer without the internet?

d How many people have a mobile phone without a camera and a computer without the internet?

8 Teachers were asked to choose their favourite snack out of chocolate and sweets.

	Chocolate	Sweets
Male	24	
Female	16	

a Copy and complete the table if 50 males were asked and 30 teachers chose sweets.

b How many females were asked?

c How many people were asked altogether?

d How many people chose chocolate?

9

Age \ Car	Porsche	BMW	Skoda
20–30	24	6	
30–40	4		14
40–50		5	40

Sam had to conduct a survey on people's age and the type of car they would like to drive.

Forty people liked the Porsche best.

Sixty people aged 30–40 were asked.

Altogether Sam asked 200 people.

Use this information to copy and complete the table.
Then, using the table, answer the following questions.

a Which is the most popular car?

b Which result do you find surprising?

c Can you think of one problem with Sam's survey?

10 The table below shows the cost of going on holiday for a week in Spain.

	May	June	July
Adult	£152	£174	£189
Child (under 16)	£120	£189	£209
Child (under 5)	Free	£25	£30

Use the table to find the cost for:

a a family of 2 adults and 1 child aged 12 to go on holiday for a week in June

b the Smith family (2 adults and 3 children, Sam 14, Jack 12 and Sally 3) to go on holiday for a week in July.

Homework 2

1 For each of the following state whether the data is quantitative or qualitative.

a Ages of people.

b Favourite film ever!

c Goals scored in a school hockey match.

d Students' favourite teacher at a school.

e The number of police at a football match.

f The time it takes to get home.

g The softness of a pile of towels.

2 For each of the following say whether the data is discrete or continuous.

  a Ages of people.

  b Goals scored in a school hockey match.

  c The number of votes in a council election.

  d The amount of water consumed by a household.

  e The viewing figures for a TV programme.

  f The time it takes to get home.

  g The number of stars that can be seen in the sky.

3 Connect each of the following to its proper description. The first one has been done for you.

The age of lions at London Zoo	**Quantitative and discrete**
People's favourite cake at a show	
Points scored in 10 darts matches	
The weight of 12 newborn puppies	**Qualitative**
The favourite building of people in Britain	
The average speed of train journeys into London	
The number of hours spent driving a day	**Quantitative and continuous**
A person's shirt collar size	
The cost of bread	
Rainfall at a seaside resort	

4 The following questions are taken from different surveys.

Write down one criticism of each question.

Rewrite the question in a more suitable form.

  a How many hours of homework do you complete each week?

  Less than 1 hour ☐ More than 1 hour ☐

  b What is your favourite colour?

  Red ☐ Blue ☐

  c How old are you?

  Under 16 ☐ Under 21 ☐ Under 40 ☐

  d You do like mathematics don't you?

  Yes ☐ No ☐

  e What is your favourite magazine?

  Hello ☐ What car? ☐

5 Write down one advantage and one disadvantage of carrying out a personal (face-to-face) survey.

6 Write down a definition and give an example of each of the following kinds of data.

a Quantitative

b Qualitative

c Continuous

d Discrete

7 Give one advantage and one disadvantage of:

a primary data

b secondary data.

Homework 3

1 Which sampling methods are most appropriate for the following surveys?

Give a reason for your answer.

a The average weight of sheep on a farm with 1000 sheep.

b The favourite building of people in your town.

c The average amount of time spent on homework each week by students in your school.

d The average hand span of students in a school.

e The views of villagers on a new shopping centre.

f Information on voting intentions at a general election.

2 The following questions are taken from different surveys.

Write down one criticism of each question and explain how you would improve it.

a A company wants to find out if people like a new bar of chocolate they have made.

They stop people in the street and ask them to try it.

They give everyone a questionnaire asking them to write everything they like about the new bar.

b A London football club wants to build a new football stadium.

They carry out a survey in central London.

They ask 1000 people if they think it will be a good idea.

c Local commuters are trying to get an improved train service into London.

They collect signatures on the train one morning.

Three quarters of the respondents say that the train service could be improved.

d Jenny is collecting information on waiting times at her local health centre.

She gets permission from the doctors to undertake her survey.

She arrives one morning and makes a note of how long each patient waits.

e You want to find out if people like motor racing.

You carry out a face-to-face survey.

You ask 100 people watching the British Grand Prix at Silverstone.

f *Strike it lucky*, a match company, want to find out the average number of matches in its boxes.

It takes a random sample of 600 boxes and counts how many matches are in each box.

It carries out the survey over a period of six weeks.

3 Write down one advantage and one disadvantage of each of the following sampling methods.

a Convenience sampling

b Random sampling

c Systematic sampling

d Quota sampling

4 A club wishes to undertake a survey of its members on whether to ban smoking on the premises.

a Explain how you would take:

i a convenience sample of 100 members

ii a random sample of 100 members.

b Which is the most appropriate sampling method? Give a reason for your answer.

5 A police force employs 200 female and 2300 male police officers. They wish to take a sample of 140 police officers to see what they do in their spare time.

Explain how you would take:

a a random sample of 140 police officers

b a systematic sample of 140 police officers.

17 Percentages

1 Change each decimal to a percentage.

   a 0.73   **b** 0.39   **c** 0.4   **d** 0.04   **e** 0.175

2 Change each percentage to a decimal.

   a 61%   **b** 22%   **c** 30%   **d** 5%   **e** $4\frac{3}{4}\%$

3 Change each fraction to a percentage.

   a $\frac{83}{100}$ **b** $\frac{1}{100}$ **c** $\frac{9}{10}$ **d** $\frac{1}{20}$ **e** $\frac{7}{8}$

4 Change each percentage to a fraction, in its simplest form.

   a 29%   **b** 6%   **c** 95%   **d** 32%   **e** $17\frac{1}{2}\%$

5 Place these in order of size, smallest first.

   a $\frac{1}{5}$, 0.18, 21%   **b** 72%, $\frac{3}{4}$, 0.7   **c** 0.06, 5%, $\frac{1}{25}$

6 Dave says that 8% is greater than 0.79 because 80 is greater than 79.
Explain his mistake.

7 Vicky changes 64% to a fraction in its simplest form.
She gets $\frac{32}{50}$. Explain why this is not the simplest form.

8 Change each of these fractions to a percentage and use
your answers to fill in the number square.

   $\frac{11}{20}$, $1\frac{1}{4}$, $\frac{9}{50}$, $\frac{18}{25}$

One answer has been put into the square to help you.

Homework 2

Work out the following:

1 10% of £70

4 30% of 80p

2 6% of 500 g

5 15% of £4.20

3 25% of 24 cm

6 5% of 60 cm

7 **Get Real!**

There are 250 students in Uptown Primary School.

24% of them take sandwiches for lunch.

How many students take sandwiches?

8 Work out the simple interest on £500 for 2 years at 7% per annum.

9 Work out the simple interest on £300 for 4 years at 6% per annum.

Work out the following:

10 55% of £3.80

13 22% of £356

11 38% of £750

14 14% of £17.28

12 90% of 13 m

15 8% of 15.4 kg

16 **Get Real!**

A restaurant has seats for 60 people.

On Tuesday evening, 35% of the seats are occupied.

How many seats are empty?

17 Work out the simple interest on £850 for 3 years at 9%.

18 Work out the simple interest on £6257 for 5 years at $8\frac{1}{2}$%.

19 Find the total owed if £4600 is borrowed for 3 years at simple interest of 9.4%.

20 Find the total owed if £9385 is borrowed for 2 years at simple interest of 10.6%.

Homework 3

 1 Increase 90p by 10%.　　 **3** Decrease 200 m by 35%.

 2 Decrease £5 by 12%.　　 **4** Increase 40 g by 15%.

 5 **Get Real!**

The bus fare to the shops is £1.50

It goes up by 8%.

What is the new fare?

 6 **Get Real!**

A plant is 16 cm high.

In one week it grows 25% higher.

What is its height at the end of the week?

 7 **Get Real!**

> **BEST USED CARS IN TOWN**
> Drive away a bargain today
> Cash discount **5%**

A car is priced at £4500.

Paula pays cash for the car and gets a 5% discount.

How much does she pay?

 8 **Get Real!**

In June, Sheds'R'Us sold 240 sheds.

In July, sales fell by 30%.

How many sheds were sold in July?

 9 Increase 350 g by 44%.　　 **11** Decrease 16 m by 7%.

 10 Increase £275 by 24%.　　 **12** Decrease £12.80 by 17.5%.

13 Get Real!

> **SPRING SALE**
> **35%** OFF ALL MARKED PRICES

A sofa had a price ticket for £1385.

In the sale, everything was reduced by 35%.

What was the sale price of the sofa?

14 Get Real!

In the same sale, Rageh bought a bookcase originally priced at £199.99

How much did he pay for it in the sale?

15 Get Real!

Sarah's salary is £13 570 per year.

She is given a bonus of $1\frac{1}{2}$% of her salary.

How much does she earn altogether?

16 Get Real!

Jack sells computers.

He is paid commission of $8\frac{1}{4}$% on his sales.

Last year he sold computers worth £85 496

How much was his commission?

Homework 4

1 Express £18 as a percentage of £300.

2 Express 45p as a percentage of 60p.

3 Express 14 kg as a percentage of 70 kg.

4 Express 3.5 m as a percentage of 14 m.

5 Get Real!

There are 30 students in the science class.

Of these 12 got full marks in a test.

What percentage of the class got full marks?

6 Get Real!

A cinema studio has 240 seats.

Of these, 12 seats have induction loops for the hard of hearing.

What percentage is this?

7 Express £81 as a percentage of £450.

8 Express 5.5 g as a percentage of 44 g.

9 Express 72p as a percentage of £1.60

10 Express 54 cm as a percentage of 2.25 m

11 Get Real!

Steve got 57 marks out of 95 for his history test.

Express this result as a percentage.

12 Get Real!

A 275 ml fruit drink contains 132 ml of fruit juice.

Express the quantity of fruit juice as a percentage of the drink.

13 Get Real!

A manager works a $37\frac{1}{2}$ hour week.

She spends 6 hours each week in finance meetings.

What percentage of her working week is this?

14 Get Real!

9875 votes were cast in an election.

Paterson got 3950 votes, Quigley got 3555 votes, Roper got 1896 votes and Stapleton got 474 votes.

Express each candidate's share of the votes as a percentage.

How can you check your answers?

Homework 5

1 The price of a packet of cereal goes up from £1.40 to £1.54

Find the percentage increase.

2   The rent on Jade's flat goes up from £80 per week to £84 per week.
Find the percentage increase.

3   In the autumn term, Sean scored 15 goals.
In the spring term he scored 21 goals.
Find the percentage increase.

4   In a spring sale, the price of a coat is reduced from £60 to £42.
Find the percentage reduction.

5   Tim bought 100 fluffy toys at 50p each.
He sold 65 of them at £1.20 each.
a How much did he pay for the toys?
b How much money did he get from selling the toys?
c Find the profit as a percentage of his costs.

6   The price of a hardback book is £15.
The paperback edition of the same book is £7.95
Express the reduction as a percentage of the hardback price.

7   In March, Anna's sales totalled £44 000.
In April her sales were £36 960.
Find the percentage decrease.

8   In a spring sale, the price of a coat is reduced from £57.50 to £41.40
Find the percentage reduction.

9   Rajesh bought a car for £8400 and sold it a year later for £5544.
Find his percentage loss.

10 Kate bought 150 antique plates for £28 each.
She sold 98 of them for £54 each and a further 24 for £35 each.
a How much did she pay for the plates?
b How much money did she get from selling the plates?
c Find the profit as a percentage of her costs.

Handling data coursework task

As part of your GCSE course you are required to submit two coursework tasks covering:

- Handling data
- Using and applying mathematics

Each coursework task is worth 10% of the total marks for your GCSE so it is important that you spend time on each piece of coursework and try and get as high a mark as possible.

This chapter explains how to get the best marks in the Handling data coursework task. The Using and applying mathematics coursework task is discussed in Homework Book 2.

Marks are awarded for the coursework task under three headings, called strands. Each strand is marked out of 8 marks. At the foundation tier, you should aim to get between 3 and 6 marks.

The criteria for the Handling data coursework are provided in pairs so 3–4 means 3 marks (if all the criteria are just met) or 4 marks (if all the criteria are confidently met).

Strand 1　Specifying the problem and planning

This strand is about choosing a statistical hypothesis and deciding what to do and how to do it. The strand requires you to provide clear aims, consider the collection of data, identify practical problems and explain how you might overcome them.

Mark	Specifying the problem and planning
3–4	Provide a clear hypothesis. Make a plan to meet the aims of the task. Your sample must be appropriate to the problem.
5–6	Provide a clear hypothesis that develops as the task proceeds. Your planning meets the aim of the task and reacts to problems with the data. Your sampling is appropriate in terms of sample size and sampling method and the reasons for this are explained.
*7–8	Provide clear aims using statistical terms. Your planning is detailed and carefully considered, meets the aims of the task and anticipates problems with the data. Your sampling is appropriate, takes account and acts upon limitations and bias is well explained. You produce a well structured report.

At the foundation tier, you should aim to get between 3 and 6 marks.
* The content for marks of 7 and 8 is included here for information.

Strand 2 Collecting, processing and representing data

This strand is about collecting data and using appropriate statistical techniques and calculations to process and represent the data. Calculations should be correct and diagrams should be appropriate and accurate (using graph paper where appropriate).

Mark	Collecting, processing and representing data
3–4	Your calculations and representations are correct and relevant. You must use several techniques and diagrams including mean, mode, median, range, pie charts, scatter graphs and stem-and-leaf diagrams.
5–6	Use your calculations and diagrams which must be correct and relevant. You must interpret a range of techniques and diagrams including mean, mode and median (of grouped data), lines of best fit, cumulative frequency diagrams and box plots.
*7–8	Your calculations and diagrams must be correct and appropriate. You must use and interpret an appropriate range of techniques and diagrams at a high level.

At the foundation tier, you should aim to get between 3 and 6 marks.
* The content for marks of 7 and 8 is included here for information.

Strand 3   Interpreting and discussing the results

This strand is about commenting, summarising and interpreting your data. Your commentary should link back to the original hypotheses and provide reflective comments on the strengths and/or weaknesses of the work.

Mark	Interpreting and discussing the results
3–4	You must comment upon your calculations and representations. Make some attempt to evaluate your method.
5–6	Calculations and representations must be commented upon and interpreted. These interpretations must be related to and linked back to the hypotheses. Provide reflective comments on the strength and/or weakness of the work.
*7–8	Calculations and representations must be summarised and correctly interpreted in detail. Overall interpretations must be related to and linked back to the hypotheses. Reflective comments must be provided recognising limitations and suggesting improvements.

At the foundation tier, you should aim to get between 3 and 6 marks.
* The content for marks of 7 and 8 is included here for information.

Explore 1   Guestimate

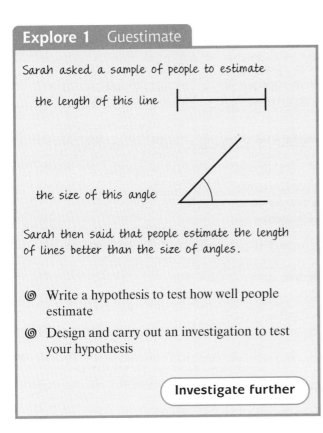

Sarah asked a sample of people to estimate

the length of this line

the size of this angle

Sarah then said that people estimate the length of lines better than the size of angles.

- ◎ Write a hypothesis to test how well people estimate
- ◎ Design and carry out an investigation to test your hypothesis

Investigate further

This task is an AQA set task so can be submitted for marking by AQA.

The words 'Investigate further' mean that you should develop the task beyond its original scope.

There are no right or wrong ways in which the task can be developed. Think of different ways in which you might extend the task.

Some tips

You may wish to consider the following:

- ◎ Think carefully about a suitable hypothesis. You do not have to consider lines and angles – these are just suggestions. Consider what information you need to know and how you can collect the information. What is an appropriate sample size? What is an appropriate sampling method?

- ◎ Consider what calculations and representations are appropriate to pursue your hypothesis. Why have you chosen these calculations and representations? Why are they appropriate? Remember to explain this.

- ◎ Remember to interpret (explain) your calculations and representations, not just describe them. Link these interpretations back to the original hypothesis. Do your findings support the hypothesis? You will need to investigate further.

- ◎ Think about how you might develop the task. What further information do you need to collect? How can you collect the information? Remember to explain what you are doing and why you are doing it.

- ◎ Remember to give an evaluation of what you have done. Reflect on the strengths and/or weaknesses of the work. Try to recognise any limitations and suggest improvements.

Explore 2   Where in the world?

Nadia is studying different countries for her geography lesson.

She thinks that there is a link between climate and tourism. She uses an atlas and the internet to find the information.

This task is an AQA set task so can be submitted for marking by AQA.

The words 'Investigate further' mean that you should develop the task beyond its original scope.

There are no right or wrong ways in which the task can be developed. Think of different ways in which you might extend the task.

◎ Write your own hypothesis linking data from a country of your choice

◎ Design and carry out an investigation to test your hypothesis

(**Investigate further**)

Some tips

You may wish to consider the following:

◎ Think carefully about a suitable hypothesis. You do not have to consider climate or tourism – these are just suggestions. Consider what information you need to know and how you can collect the information.

◎ Will the data be primary data (data you collect yourself) or secondary data (data that has already been collected by someone else)? Where will you find this data?

◎ What is an appropriate sample size? What is an appropriate sampling method? Why are you doing it in this way? What calculations and representations are appropriate to pursue your hypothesis? Remember to explain this.

◎ Remember to interpret (explain) your calculations and representations, not just describe them. Link these interpretations back to the original hypothesis. Do your findings support the hypothesis? You will need to investigate further.

◎ Think about how you might develop the task. What further information do you need to collect and is this information readily available? Remember to explain what you are doing and why you are doing it.

◎ At the end of the work remember to provide an evaluation of what you have done. Reflect on the strengths and/or weaknesses of the work. Try to recognise any limitations and suggest improvements.

Explore 3 Body parts

Leonardo da Vinci's famous drawing called the Vitruvian Man shows a picture of a man with outstretched arms and legs.

The picture suggests that the length of a man's arm span is equal to the height of a man.

This task is not an AQA set task so must be marked by the centre and moderated by AQA.

The words 'Investigate further' mean that you should develop the task beyond its original scope.

There are no right or wrong ways in which the task can be developed. Think of different ways in which you might extend the task.

- ⊚ Write a hypothesis to test this assumption
- ⊚ Design and carry out an investigation to test your hypothesis

Investigate further

Some tips

You may wish to consider the following:

- ⊚ Think carefully about a suitable hypothesis. You do not have to consider the length of a man's arm span and height – these are just suggestions. Consider what information you need to know and how you can collect the information. What is an appropriate sample size? What is an appropriate sampling method?

- ⊚ Will the data be primary data (data you collect yourself) or secondary data (data that has already been collected by someone else)? How will you collect this data or where will you find this data?

- ⊚ What calculations and representations are appropriate to pursue your hypothesis? Are the calculations and representations appropriate? Remember to explain this.

- ⊚ Remember to interpret (explain) your calculations and representations, not just describe them. Link these interpretations back to the original hypothesis. Do your findings support the hypothesis? You may need to investigate further.

- ⊚ Think about how you might develop the task. What further information do you need to collect and is this information readily available? Remember to explain what you are doing and why you are doing it.

- ⊚ At the end of the work remember to provide an evaluation of what you have done. Reflect on the strengths and/or weaknesses of the work and try to recognise any limitations and suggest improvements.